# Springer Theses

Recognizing Outstanding Ph.D. Research

## Aims and Scope

The series "Springer Theses" brings together a selection of the very best Ph.D. theses from around the world and across the physical sciences. Each volume has been selected for its scientific excellence by an internationally renowned institute or department. For greater accessibility to non-specialists, the published versions include an extended introduction, as well as a foreword by the student's supervisor explaining the special relevance of the work for the field. The series as a whole provides a valuable resource both for newcomers to the research fields, and for other scientists seeking detailed background information on special questions. Finally, it serves as an accredited documentation of the valuable contributions made by today's younger generation of scientists.

## How to Apply

Theses prepared at internationally leading universities and institutes may be nominated for publication by supervisors with the endorsement of the head of department. Each such department is entitled to nominate one thesis per year. Enquiries and requests for a nomination form should be sent, together with the thesis, to Ms. Nathalie Berg (see link on right).

## Eligibility Criteria

To qualify for consideration, theses should fulfil the following criteria:

- They must be written in good English.
- The topic should fall within the confines of Chemistry, Physics, Earth Sciences, Engineering and related interdisciplinary fields such as Materials, Nanoscience, Chemical Engineering, Complex Systems and Biophysics.
- The work reported in the thesis must represent a significant scientific advance.
- A maximum of 30% of the text may be a verbatim reproduction from the author's previous publications.
- They must have been examined and passed during the 12 months prior to nomination.
- The theses should have a clearly defined structure including an introduction accessible to new PhD students and scientists not expert in the relevant field.
- After acceptance, theses should be supplemented by a "Supervisor's Foreword" outlining the wider scientific relevance of the work

Indexed by zbMATH.

Yuichi Sakuda

# Discovery of $Ba_7Nb_4MoO_{20}$-Based Materials and the Mechanism of Ultrafast Ion Conduction Via Dimers

Doctoral Thesis accepted by
Tokyo Institute of Technology, Tokyo, Japan

Springer

*Author*
Dr. Yuichi Sakuda
Department of Chemistry
Tokyo Institute of Technology
Tokyo, Japan

*Supervisor*
Prof. Masatomo Yashima
Institute of Science Tokyo
Tokyo, Japan

ISSN 2190-5053 ISSN 2190-5061 (electronic)
Springer Theses
ISBN 978-981-96-2551-2 ISBN 978-981-96-2552-9 (eBook)
https://doi.org/10.1007/978-981-96-2552-9



This Springer imprint is published by the registered company Springer Nature Singapore Pte Ltd.
The registered company address is: 152 Beach Road, #21-01/04 Gateway East, Singapore 189721, Singapore

# Supervisor's Foreword

I highly recommend this book of Sakuda's Ph.D. research. This work focused on the both the discovery of the high ion conductor and the elucidation of the ion migration mechanism in the $Ba_7Nb_4MoO_{20}$-based hexagonal perovskite-related oxides, which were investigated by experiments and theoretical calculations.

First, the $Cr^{6+}$ doping into $Ba_7Nb_4MoO_{20}$ simultaneously improved the oxide-ion conductivity and suppressed the proton conductivity. Neutron diffraction data and Rietveld analysis showed that the $Cr^{6+}$ cation is located at the Nb/Mo2 site near the intrinsic oxygen-deficient layer. Maximum entropy method (MEM) analysis showed that oxide ions diffuse through the lattice O1 and interstitial O5 sites in the intrinsic oxygen deficient layer via the interstitialcy diffusion mechanism.

Second, the introduction of excess oxygen atoms by the Mo/Nb substitution in $Ba_7Nb_4MoO_{20}$ improved oxide-ion and proton conductivity. $Ba_7Nb_{3.8}Mo_{1.2}O_{20.1}$ exhibited high oxide-ion and dual-ion (oxide-ion and proton) conductivity. The oxide-ion conductivity of $Ba_7Nb_{3.8}Mo_{1.2}O_{20.1}$ was 175 times higher than that of conventional yttria-stabilized zirconia (YSZ) at 306 °C. The origin of the high oxide-ion conductivity at low temperature for $Ba_7Nb_{3.8}Mo_{1.2}O_{20.1}$ was found to be the low activation energy for conductivity due to the long (Nb/Mo2)–Ba2 distance sandwiching the oxide-ion conducting layer. Ab initio molecular dynamics (AIMD) simulation of ~ $(Ba_7Nb_{3.8}Mo_{1.2}O_{20.1})_9$ showed that oxide ions migrate by concerted migration via the breaking and reforming of $(Nb/Mo2)_2O_9$ dimers, $(Nb/Mo2)O_5$ monomers, and $(Nb/Mo2)O_4$ tetrahedra. The AIMD simulations of ~ $(Ba_7Nb_{3.8}Mo_{1.2}O_{20.1}\cdot 0.1\ H_2O)_9$ showed that proton diffusion occurs mainly via the long-range proton migration in the oxygen close-packed layer in $Ba_7Nb_{3.8}Mo_{1.2}O_{20.1}$.

Overall, this work is very interesting in terms of both controlling the ion conductivity by element substitution and elucidating the ion migration pathway in hexagonal perovskite-related oxides. Furthermore, this research has provided a guideline for the search for new ion conductors and has made a significant contribution to the field of ion conductors.

Tokyo, Japan
November 2024

Masatomo Yashima

# Acknowledgements

I would like to take this opportunity to express my sincere gratitude to the many people who helped me in carrying out this research and compiling this thesis. In particular, my supervisor Prof. Masatomo Yashima has been very helpful in many areas, such as by providing daily research guidance, helping me with writing papers during meetings, checking my progress through seminars, and consulting with me about the direction of my future research.

I would like to express my sincere gratitude to Profs. Shinya Koshihara, Koji Taniguchi, Masaki Kawano, and Associate Professor Hidehiro Uekusa for their guidance in reviewing this thesis.

Assistant Professor Kotaro Fujii provided us with advice on crystal structure analysis and calculations, as well as assistance with measurement experiments at off-campus facilities. Special Assistant Professor Taito Murakami provided us with various guidance in writing papers and applications. We would like to express our gratitude to both of them.

I would like to express our gratitude to Dr. Mudasir A. Yatoo and Prof. Stephen J. Skinner of Imperial College London for providing us with the opportunity to conduct diffusion experiments and hold many discussions as part of our collaborative research.

I would like to thank Dr. Maxim Avdeev of ANSTO for his help with the AIMD calculations, Dr. James R. Hester of ANSTO for his help with the neutron diffraction measurements, Dr. Masato Hagihala of the High Energy Accelerator Research Institute, Dr. Yoichi Ikeda and Dr. Yusuke Nanbu of Tohoku University, Mr. Katsumi Suda of the Analysis Division of the Open Facility Center at Tokyo Institute of Technology for his help with the TG-MS measurements, Dr. Masaru Tada of the Analysis Division of the Open Facility Center at Tokyo Institute of Technology for his help with the Raman spectrum measurements, Dr. Koji Chiba and Researcher Junko Habasaki of MOLSYS Co., Ltd. for their extensive advice on the calculations, Dr. Fujio Izumi and Dr. Koichi Momma of the Institute for Materials Science, National Institute for Materials Science, and Dr. Takashi Kamiyama, Dr. Ryoko Tomiyasu, and Dr. Masao Yonemura of the High Energy Accelerator Research Organization, for providing us with the software used in the analyses in this study.

I would like to thank for the reproduced with permission from the Royal Society of Chemistry, American Chemical Society, Springer Nature and Journal of the Ceramic Society of Japan.

I would like to express my sincere gratitude to all the members of the Yashima Laboratory, including my seniors who have graduated, for their help in many areas, such as analysis and experiments in the lab, various discussions through seminars, research presentations outside the lab, and experimental facilities.

# Contents

# Chapter 1
# Introduction

## 1.1 Ion Conductors

Recently, we have been facing various environmental problems including global warming. Since one of the causes of global warming is the emission of greenhouse gases such as $CO_2$, it is important to promote power generation that does not emit greenhouse gases [1]. Fuel cells are an environmentally friendly method of generating electricity by producing water from hydrogen and oxygen. It is hoped that switching to fuel cell-based power generation and improving power generation efficiency will help slow global warming. Fuel cells consist of an air electrode, an electrolyte, and an anode, and there are several types of fuel cells. The two most efficient fuel cells are solid oxide fuel cells (SOFCs) and proton conducting ceramic fuel cells (PCFCs). The electrolyte used in SOFCs is an "oxide ion conductor", in which oxide ions move through the solid, while the electrolyte used in PCFCs is a "proton conductor", in which protons move [2, 3]. Therefore, the discovery of materials that exhibit high ion conductivity will promote the development of SOFCs and PCFCs, and provide a clue to solving the problem of global warming. Since Tubandt et al. discovered that $\alpha$-AgI exhibits high ionic conductivity in 1914 [4], many compounds that conduct ions with a valence of + 2 to – 2 have been reported to date, including lithium-ion conductors [5], oxide-ion conductors [6–11], proton conductors [12, 13], sodium-ion conductors [14, 15], copper-ion conductors [16], silver-ion conductors [17], fluoride-ion conductors [18], and magnesium-ion conductors [19, 20].

## 1.2 Oxide-Ion Conductors

To date, many structures have been reported for oxide-ion conductors, including yttria-stabilized zirconia (YSZ), which has a fluorite structure and is a practical material [21–24]. For example, $ABX_3$ type perovskite structure [6, 25–29], Dion-Jacobson

 
Y. Sakuda, *Discovery of $Ba_7Nb_4MoO_{20}$-Based Materials and the Mechanism of Ultrafast Ion Conduction via Dimers*, Springer Theses,
https://doi.org/10.1007/978-981-96-2552-9_1

type structure [30, 31], $K_2NiF_4$ type structure [32–34], apatite type structure [35–38], melilite type structure [39–42], LAMOX type structure [43–45], scheelite type structure [46–51], pyrochlore type structure [52], $BaNdInO_4$ type structure [53–55], and hexagonal perovskite-related structures [7, 56–71] are listed. Furthermore, strategies have been developed to improve oxide-ion conductivity by changing the carrier concentration through changing the amount of oxygen in the crystal structure through element substitution and doping. The first strategy is to increase the number of carriers by creating oxygen vacancies by removing some of the oxygen present in the lattice sites. For example, regarding the practical material YSZ, partial substitution of $Zr^{4+}$ with $Y^{3+}$ introduces oxygen vacancy through charge compensation, and the material exhibits oxide-ion conductivity exceeding $10^{-2}$ S cm$^{-1}$ at 700 °C (Fig. 1.1) [23]. The other strategy is to introduce oxygen into interstitial sites to increase the number of carriers. For example, regarding the melilite-type structure $LaSrGa_3O_7$, partial substitution of $Sr^{2+}$ with $La^{3+}$ introduces interstitial oxygen through charge compensation, and the oxide-ion conductivity exceeds $10^{-1}$ S cm$^{-1}$ at 1000 °C. [39] From the above, in order to discover a material with high oxide-ion conductivity, it is important not only to discover a structure type that exhibits high oxide-ion conductivity, but also to control the amount of oxygen by doping. The problem with the electrolyte materials currently in practical use for SOFCs is that YSZ has a high activation energy for oxide-ion conductivity, and its oxide-ion conductivity at low temperatures is insufficient, making it difficult to put it to practical use at low temperatures. Therefore, in this study, I aim to discover the materials that exceed the oxide-ion conductivity of YSZ and other materials at low temperatures, in order to get closer to practical use of SOFCs that operate at low temperatures (Fig. 1.1).

## 1.3 Proton Conductors

As with oxide-ion conductors, several structures that exhibit proton conductivity have been reported [77–80]. For example, the $ABX_3$-type perovskite structure [81–83], the scheelite-type structure [84, 85], the brownmillerite-type structure [86], the $BaNdGaO_4$-type structure [87], the apatite-type structure [88], and hexagonal perovskite-related structures [63–65, 89–94] are listed. A common strategy to improve proton conductivity is to create oxygen vacancies in the crystal structure by element substitution/doping, introduce $OH^-$ into the vacancies, and increase the carrier concentration. For example, in $BaZrO_3$, partial substitution of $Zr^{4+}$ with $Y^{3+}$ introduces oxygen vacancy through charge compensation, and the introduction of $OH^-$ shows proton conductivity exceeding $10^{-2}$ S cm$^{-1}$ at 500 °C [95]. Proton conductivity $\sigma_{H+}$ is expressed by the formula $\sigma_{H+} = n_H \times \mu \times Ze$ ($n_H$: carrier concentration, $\mu$: mobility, Ze: charge), so in order to improve $\sigma_{H+}$, it is important to increase the carrier concentration, i.e., to increase the amount of protons.

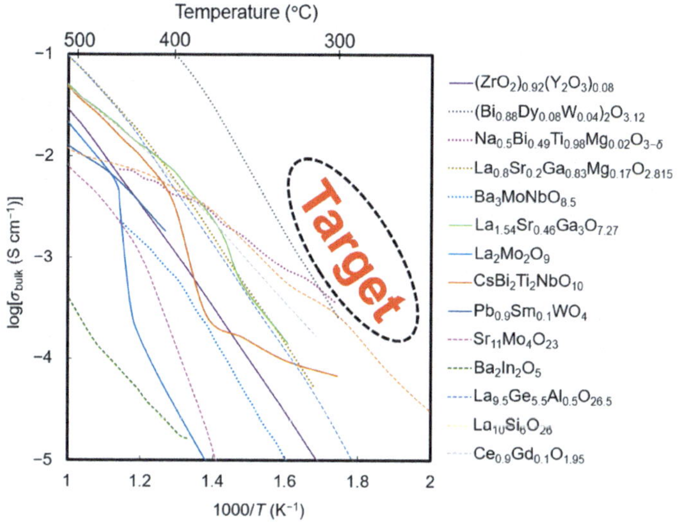


**Fig. 1.1** Bulk conductivity of representative oxide-ion conductors ($(ZrO_2)_{0.92}(Y_2O_3)_{0.08}$, [23] $(Bi_{0.88}Dy_{0.08}W_{0.04})_2O_{3.12}$, [24] $Na_{0.5}Bi_{0.49}Ti_{0.98}Mg_{0.02}O_{3-\delta}$, [72] $La_{0.8}Sr_{0.2}Ga_{0.83}Mg_{0.17}O_{2.815}$, [26] $Ba_3MoNbO_{8.5}$, [58] $La_{1.54}Sr_{0.46}Ga_3O_{7.27}$, [39] $La_2Mo_2O_9$, [73] $CsBi_2Ti_2NbO_{10}$, [31] $Pb_{0.9}Sm_{0.1}WO_4$, [47] $Sr_{11}Mo_4O_{23}$, [74] $Ba_2In_2O_5$, [75] $La_{9.5}Ge_{5.5}Al_{0.5}O_{26.5}$, [76] $La_{10}Si_6O_{26}$, [35] $Ce_{0.9}Gd_{0.1}O_{1.95}$ [21]) and target material

## 1.4 Dual-Ion Conductors

A dual ion conductor is a material that exhibits both oxide-ion conductivity and proton conductivity. The advantages of using a dual-ion conductor as an electrolyte are as follows; (i) Oxide-ions and protons migrate simultaneously in the electrolyte, improving the total electrical conductivity, which is expected to improve cell performance [96]. (ii) Humidification by an external gas is not required, greatly simplifying the fuel cell system [97]. Recently, $BaZr_{0.1}Ce_{0.7}Y_{0.1}Yb_{0.1}O_{3-\delta}$ (BZCYYb) is attracting attention as a dual-ion conductor, and it shows high dual-ion conductivity exceeding $10^{-2}$ S cm$^{-1}$ at 450 °C (Fig. 1.2) [98]. Therefore, in this work, I aim to discover a new dual-ion conductor that exceeds the dual-ion conductivity of BZCYYb. Furthermore, regarding fuel cells that use a dual ion conductor as the electrolyte, the cathode material requires a triple conductor (oxide ion conduction, proton conduction and electron conduction). Recently, triple conductor materials such as $PbBa_{0.5}Sr_{0.5}Co_{1.5}Fe_{0.5}O_{5+\delta}$ have been discovered [99, 100], and it is expected that dual ion type fuel cells will be put into practical use in future. However, as mentioned in Sect. 1.3, in order to exhibit proton conductivity, $OH^-$ must be introduced, which requires the creation of oxygen vacancies by doping. It is difficult to find a material that exhibits high dual-ion conductivity by the strategy of doping to increase carriers by creating interstitial oxygen.

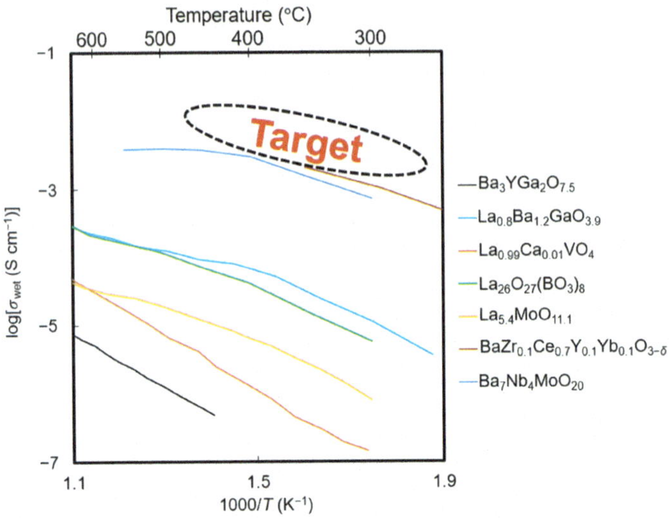


**Fig. 1.2** Conductivity of representative dual-ion conductors ($Ba_3YGa_2O_{7.5}$, [101] $La_{0.8}Ba_{1.2}GaO_{3.9}$, [102] $La_{0.99}Ca_{0.01}VO_4$, [85] $La_{26}O_{27}(BO_3)_8$, [103] $La_{5.4}MoO_{11.1}$, [104] $BaZr_{0.1}Ce_{0.7}Y_{0.1}Yb_{0.1}O_{3-\delta}$, [82] $Ba_7Nb_4MoO_{20}$ [62]) and target material in wet air

## 1.5 Ion Migration Mechanism

In Sect. 1.1, I listed the types of ionic conductors, and there have been many reports on the mechanisms by which ions migrate in crystals [10, 19, 22, 93, 94, 105–120]. In order to clarify the mechanism of ion transfer, it is important to conduct multiple experiments and calculations and consider them comprehensively. Specifically, the crystal structure of a sample can be investigated by experimental techniques such as Rietveld analysis using synchrotron radiation and neutron diffraction, maximum entropy method (MEM), quasi-elastic scattering measurement, NMR measurement, and spectroscopy. In addition, experimental results can be further supported by combining computational methods such as classical molecular dynamics and Ab initio molecular dynamics (AIMD). By the above techniques, the crystallographic parameters of mobile ions and their coordinated ions, neutron scattering length density (NSLD) distributions, and ion migration in local dynamic structures have been considered. Three mechanisms are particularly important in considering ion conduction; (i) vacancy mechanism, (ii) direct interstitial mechanism, and (iii) interstitialcy mechanism (Fig. 1.3) [121]. (i) Vacancy mechanism: some of the ions in the lattice sites become deficient, allowing ions to move through these vacancies. (ii) Direct interstitial mechanism; Ions exist between lattice sites and ions move directly from interstitial sites to interstitial sites. (iii) Interstitialcy mechanism. The ions move from lattice sites toward interstitial sites (or vice versa), and the displaced interstitial ions displace the lattice site ions, repeatedly. In addition to the three typical mechanisms (i)-(iii), more detail ion migration mechanisms have been considered, and their relationship to the transport properties of ions has been investigated.

## 1.6 Oxide-Ion Migration Mechanism

As mentioned in Sect. 1.5, there are many reports on the mechanism by which oxide ions migrate. In addition, since oxide ions need to pass between cations $M$ to move, there are also examples that focus on polyhedra $M_xO_n$ ($x$ and $n$ are natural numbers) and investigate the coordination number of $M$. A typical example is the mechanism of oxide-ion migration in $BaNdGaO_4$-type structure $La_{0.8}Ba_{1.2}GaO_{3.9}$. [87] For $LaBaGaO_4$, partial substitution of $Ba^{2+}$ with $La^{3+}$ introduces interstitial oxygen through charge compensation, and oxide ions migrate by the vacancy mechanism. In 2007, by molecular dynamics (MD), Kendrick et al. reported that oxide ions migrate by the breaking and re-forming of $Ga_2O_7$ units in snapshots (Fig. 1.4). The presence of $Ga_2O_7$ units was also confirmed by Fourier maps using neutron diffraction, which showed that $GaO_4$ tetrahedra rotated by the oxide ion trajectories (Fig. 1.5). Kendrick et al. termed this motion of oxide ions concerted "cog-wheel"-type motion. The $M_2O_n$ unit (commonly called "dimer") consisting of two cations $M$ and oxygen, breaking and re-forming, and rotation of polyhedra are examples commonly seen in compounds in which oxide ions migrate by the vacancy mechanism, and have been reported in $Sr_{0.0625}Bi_{0.9375}VO_{3.96875}$ [50] and $K_2ZnV_2O_7$, [122] etc., and are important concepts in the mechanism of oxide ion movement. However, by many MD simulations, although the existence of dimers in the local dynamic structures has been shown it is difficult to prove it experimentally.

An example of oxide ion migration by the interstitialcy mechanism is the $K_2NiF_4$-type structure $A_2BO_4$-based oxide. By Rietveld analysis and MEM analysis using neutron diffraction, several examples have been reported in which the interstitial mechanism was shown by the connected neutron scattering length density between the lattice oxygen and interstitial oxygen [33, 123, 124]. Furthermore, DFT calculations of the $A_2BO_4$-based oxide $Pr_{40}Ni_{15}Cu_4GaO_{86}$ showed that the $BO_6$ polyhedron are locally formed into $BO_7$ polyhedron by the interstitial oxygen [34]. Examples of an increase in the coordination number of $M_xO_n$ polyhedra due to the coordination of interstitial oxygen are also seen in the scheelite-type [125] and melilite-type structures [39], which show high oxide-ion conductivity. Therefore, it is important to pay attention to the change in the coordination number of cations in the polyhedra when considering the ion migration mechanism and ion conductivity.

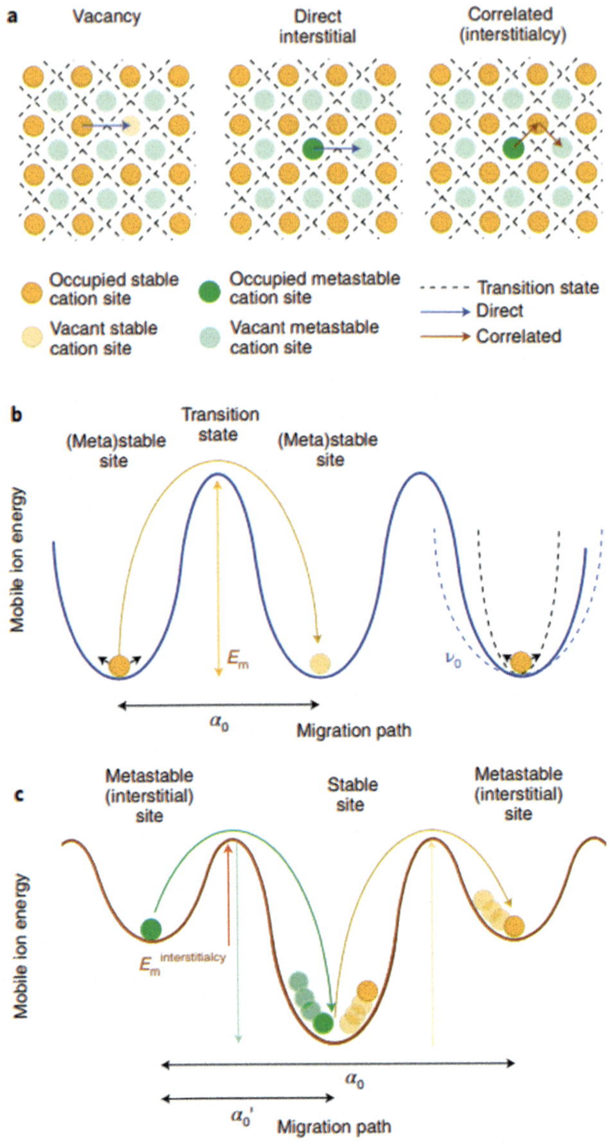


**Fig. 1.3** Cation migration mechanisms and associated energy profiles. **a** The arrows indicate three typical migration mechanisms: vacancy, direct interstitial and correlated (interstitialcy) involving a single or multiple sites (blue and red, respectively). Circles represent cations in stable (green) and metastable (orange) sites of a model crystal lattice. Dotted lines represent the transition state for cation hopping as imposed by the anionic framework (not shown explicitly). **b**, **c** The energy profiles associated with cation migration via direct vacancy or interstitial hopping (**b**) and correlated hopping (**c**) are shown with their associated hopping energies, $E_m$, hopping distances $a_0$ and hopping frequencies, $\nu_0$. [121] Copyright© 2019 Springer Nature

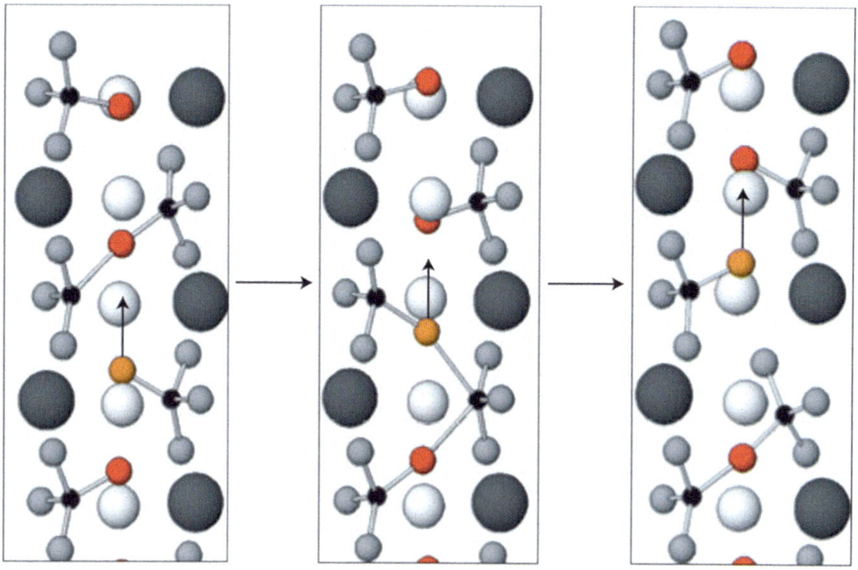

**Fig. 1.4** Lowest-energy pathway for oxygen vacancy migration in $La_{1-x}Ba_{1+x}GaO_{4-x/2}$. Snapshots taken from simulations show the microscopic mechanism of oxide-ion migration involving the formation and breaking of $Ga_2O_7$ units. The transfer of oxide ions between neighboring tetrahedra enables long-range diffusion to take place. The migrating oxide ions are highlighted, and the "stick" bonds are drawn to aid the interpretation of the transport mechanism [87]. Copyright© 2007 Springer Nature

## 1.7 Proton Migration Mechanism

Protons migrate by hopping around oxygen. Therefore, it is important to consider the migration mechanism focusing on polyhedral $MO_n$. In order for protons to diffuse, protons must migrate not only from oxygen to oxygen within a single polyhedron, but also from oxygen to oxygen between polyhedron. The former migration is intra-polyhedral, and the latter migration is inter-polyhedral (Fig. 1.6) [80]. In the simple structure of the $ABX_3$ perovskite, there is only one type of migration between $MO_6$ polyhedra, but as the number of elements increases and the structure becomes more complex, multiple migration mechanisms are possible. For example, there are layered perovskite structures such as the Dion-Jacobson structure and the (110) layered perovskite structure, which have been reported to have proton conduction. Shiraiwa et al. reported the proton migration pathways in $BaNd_{0.8}Ca_{0.2}ScO_{3.9}$ having a layered perovskite-type (110) structure by DFT calculations of the proton migration energy in and between layers [80]. In the case of a layered structure such as $BaNd_{0.8}Ca_{0.2}ScO_{3.9}$, where many polyhedra exist within the unit cell, it is common to consider the migration mechanism for each layer separately.

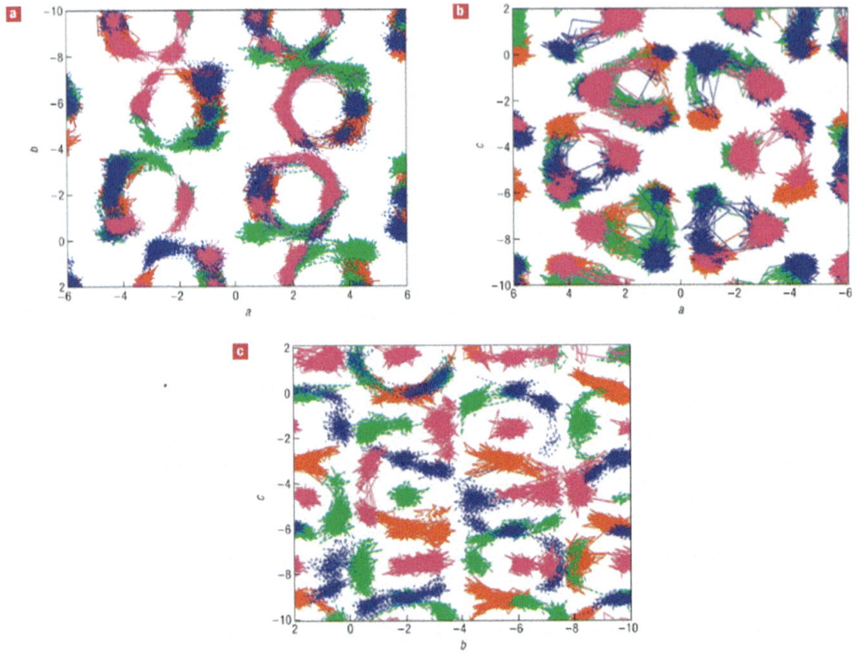


**Fig. 1.5** Molecular dynamics simulation trajectories of oxide ions in three mutually perpendicular planes. **a** Motion of the anions (red = O1, green = O2, blue = O3, pink = O4) during molecular dynamics simulations reveals that motion in the ab plane is mostly the facile rotation of the tetrahedra, and does not contribute significantly to long-range oxide-ion diffusion. **b**, **c** Motion in the ac (**b**) and bc (**c**) planes reveal more correlated diffusional features: the interconnected trajectories across the bc (**c**) plane reveals that long-range diffusion mostly takes place in this plane, indicating anisotropic migration pathways [87]. Copyright© 2007 Springer Nature

## 1.8 Hexagonal Perovskite-Related Structure and Tetrahedra

Hexagonal perovskite-related structure is one of the perovskite-type structures represented by $ABX_3$ ($A$: relatively large cation, $B$: relatively small cation, $X$: anion) [126]. Close-packed $AX_3$ layers consisting of cations $A$ and anions $X$ with similar ionic radii are stacked, and various structures have been reported depending on the way they are stacked. There are two types of close-packed structures: a cubic close-packed structure stacked in the pattern ABABAB... and a hexagonal close-packed structure stacked in the pattern ABCABCABC.... Hexagonal perovskite-related structures include hexagonal close-packed structures and/or cubic close-packed structures. There are several notation methods for hexagonal perovskite-related structures: c–h notation, ABC notation, Zhdanov notation, and Ramsdell notation. Here, I will explain the commonly used c–h notation. The c–h notation is a way to describe the positional relationship between the layers above and below the close-packed

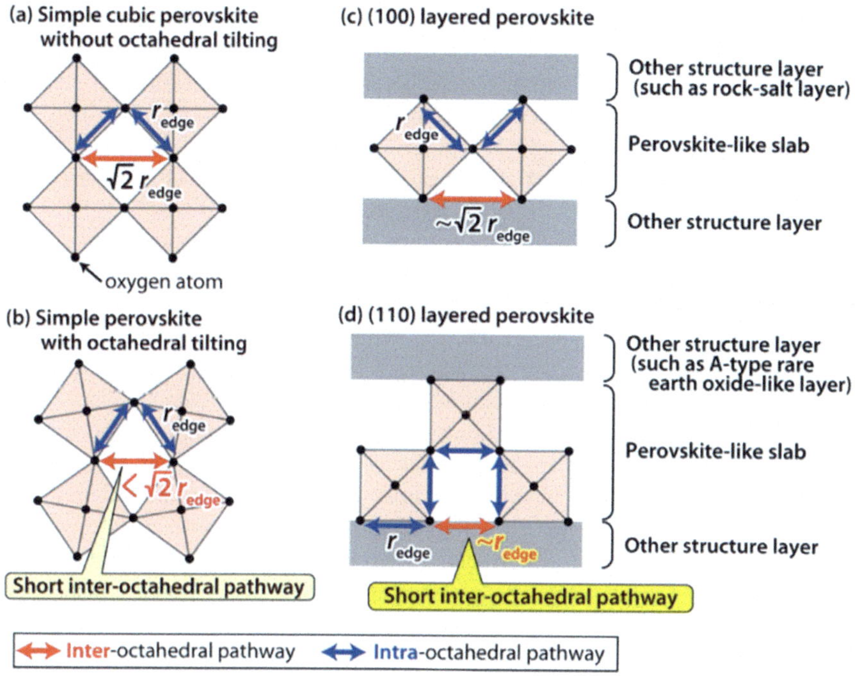


**Fig. 1.6** Schematic of inter-octahedral and intra-octahedral proton hopping in **a** a simple cubic perovskite without octahedral tilting, **b** a simple perovskite with octahedral tilting, **c** a (100) layered perovskite without octahedral tilting, and **d** a (110) layered perovskite. The black spheres and orange squares stand for the oxygen atoms and the $MX_6$ octahedron of the $AMX_3$ perovskite unit, respectively. Here, $A$ and $M$ are relatively larger and smaller cations, respectively, and $X$ is an anion. $r_{edge}$ denotes the O–O edge distance of the octahedron, which is equal to the intra-octahedral distance. The cation is omitted for simplicity [80]. Copyright© 2021 The Royal Society of Chemistry

$AX_3$ layer. When a layer A is sandwiched between layers of the same type, such as BAB or CAC, it is written as h (h stands for hexagonal). When a layer A is sandwiched between different layers, such as BAC or BAC, it is written as c (c stands for cubic). Furthermore, some hexagonal perovskite-related structures contain layers with partial anion deficiency, $AX_{3-\delta}$ ($\delta$ is the amount of anion deficiency), and such layers are represented by the symbol ′ (prime). For example, a layer sandwiched between the same layers is represented as h′, and a layer sandwiched between different layers is represented as c′. These layers have oxygen vacancies in the parent material without elemental substitution, and are so-called "intrinsically oxygen-deficient layers".

As mentioned above, the hexagonal perovskite-related structure is one of the perovskite-type structures. Although several perovskite-type structures have been reported to exhibit high ion conductivity, the hexagonal perovskite-related structure has remained largely unexplored as an ion conductor. In recent years, Since Fop et al. reported $Ba_3MoNbO_{8.5}$ in 2016 [58] and $Ba_7Nb_4MoO_{20}$ in 2020 [62], it has gradually attracted attention as a structural type that exhibits high oxide ion conductivity. Furthermore, Murakami et al. reported $Ba_5Er_2Al_2ZrO_{13}$ in 2020 [92] and $\beta$-$Ba_2ScAlO_5$ in 2022 [94] as materials that exhibit high proton conductivity, and it is gradually attracting attention as a material that exhibits high proton conductivity. These materials are characterized by the presence of an intrinsically oxygen-deficient layer. $Ba_3MoNbO_{8.5}$-based materials and $Ba_7Nb_4MoO_{20}$-based materials with the c′ layer, in which oxide ions are diffused. $Ba_5Er_2Al_2ZrO_{13}$ and $\beta$-$Ba_2ScAlO_5$ have the h′ layer. The large amount of protons amount of protons are introduced as career in the h′ layer, so the intrinsically oxygen-deficient layer has a significant effect on ion conduction. In crystal structures consisting only of close-packed oxygen layers and $B$-site cations, only $BO_6$ octahedra exist, whereas compounds containing essentially oxygen-deficient layers are characterized by the presence of $BO_4$ tetrahedra as well. Before 2000, study was focused on fluorite-type structures consisting of 8-coordinated polyhedra and perovskite-type structures consisting of 6-coordinated polyhedra. Recently, however, many materials containing tetrahedra which exhibiting high oxide-ion conductivity have been reported. In a review published in 2023, Yang et al. reported that compared with the hexa- and octa-coordinated polyhedral networks of perovskite or fluorite structures, tetrahedral-based structures have lower density, which may be advantageous for accommodating extra oxygen atoms, and that deformation of the tetrahedrons may provide more room for chemical bonding with the oxygen atoms [127]. Since hexagonal perovskite-related oxides with the intrinsically oxygen-deficient layer have vacancies for introducing oxygen, and therefore have an advantageous structure for introducing excess oxygen. In fact, when excess oxygen is introduced into the interstitial sites of the intrinsically oxygen-deficient layer in $Ba_3MoNbO_{8.5}$-based materials, it coordinates with the B-site cations as octahedral oxygen, and the large amount of tetrahedral and octahedral oxygen leads to high oxide ion conductivity [58, 60]. In 2022, Yasui et al. proposed a disorder parameter (DP) for the $Ba_3MNbO_{8.5}$ ($M$; Mo, W) that focuses on the number of tetrahedral oxygen atoms and the number of octahedral oxygen atoms in the $(M/Nb)O_n$ ($n = 4, 6$) polyhedron [69]. It was shown that the closer the number of tetrahedral oxygen atoms is to the number of octahedral oxygen atoms, the larger the DP becomes, and that as the DP increases, the minimum NSLD on the path along where the oxide ions diffuse becomes larger, thereby improving the oxide-ion conductivity of $Ba_3MNbO_{8.5}$. Based on the above, it is believed that focusing on compounds having the intrinsically oxygen-deficient layer and introducing excess oxygen into interstitial sites will be effective in discovering materials that exhibit high oxide-ion conductivity.

As mentioned in Sect. 1.6, for oxides containing tetrahedra in crystal structure, there are many reports that oxide ions migrate by breaking and re-forming of dimers due to the changing in the amount of oxygen such as the $LaBaGaO_4$ structure [87] and the scheelite structure [50, 125]. Hexagonal perovskite-related oxides with the intrinsically oxygen-deficient layers also contain $MO_4$ tetrahedral units in the parent material, so it is expected that dimers will be produced by introducing excess oxygen.

## 1.9 $Ba_7Nb_4MoO_{20}$-Based Materials

As mentioned in Sect. 1.8, $Ba_7Nb_4MoO_{20}$-based materials exhibit high oxide-ion conductivity. Figure 1.7 shows the crystal structure $Ba_7Nb_4MoO_{20}$. The crystal structure in $Ba_7Nb_4MoO_{20}$ consists of palmierite-like layers (P in Fig. 1.7a) and 12R perovskite layers (12R in Fig. 1.7a), and the intrinsically oxygen-deficient layer (c′ layer) exists in the palmierite-like layer.

In $Ba_7Nb_4MoO_{20}$-based materials, the oxide ion conductivity is improved by substituting part of the $Nb^{5+}$ with $W^{6+}$ or $Mo^{6+}$. MEM NSLD showed that in the crystal structure of $Ba_7Nb_{3.9}Mo_{1.1}O_{20.05}$ [63] and $Ba_7Nb_{3.85}W_{0.15}MoO_{20.075}$, [68] oxide ions diffuse 2-dimensionally in the c′ layer by interstitialcy mechanism between lattice oxygen O1 and interstitial oxygen O5 (Fig. 1.8), and that the improvement in oxide-ion conductivity is due to the introducing excess oxygen as interstitial oxygen.

$Ba_7Nb_4MoO_{20}$-based materials show high oxide ion conductivity compared to conventional ion conductors, making them promising electrolyte materials for SOFCs, but proton conductivity is not desirable for use as an electrolyte material. Therefore, in order to simultaneously improve oxide ion conductivity and suppress

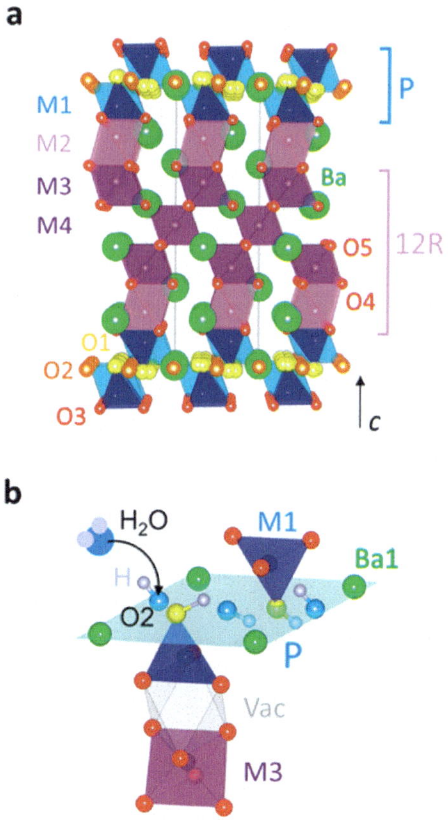


**Fig. 1.7** **a** Average crystal structure of $Ba_7Nb_4MoO_{20}$ composed of an intergrowth of palmierite-like layers (P) and 12R hexagonal perovskite blocks. Blue and light blue polyhedra represent the variable geometry $M1O_x$ units with average tetrahedral and octahedral coordination created by partial occupation of the two crystallographic oxygen positions O1 and O2. The M1 and M2 metal positions are also partially occupied, thus resulting in hybrid trimer stackings of cationic vacancies and isolated/face-sharing polyhedral units. **b** Representation of the water absorption on the vacant O2 site along the cubic $[Ba_1O_2]$ plane of the palmierite-like layer. Vac indicates a vacant M2 site [64]. Copyright© 2021 American Chemical Society

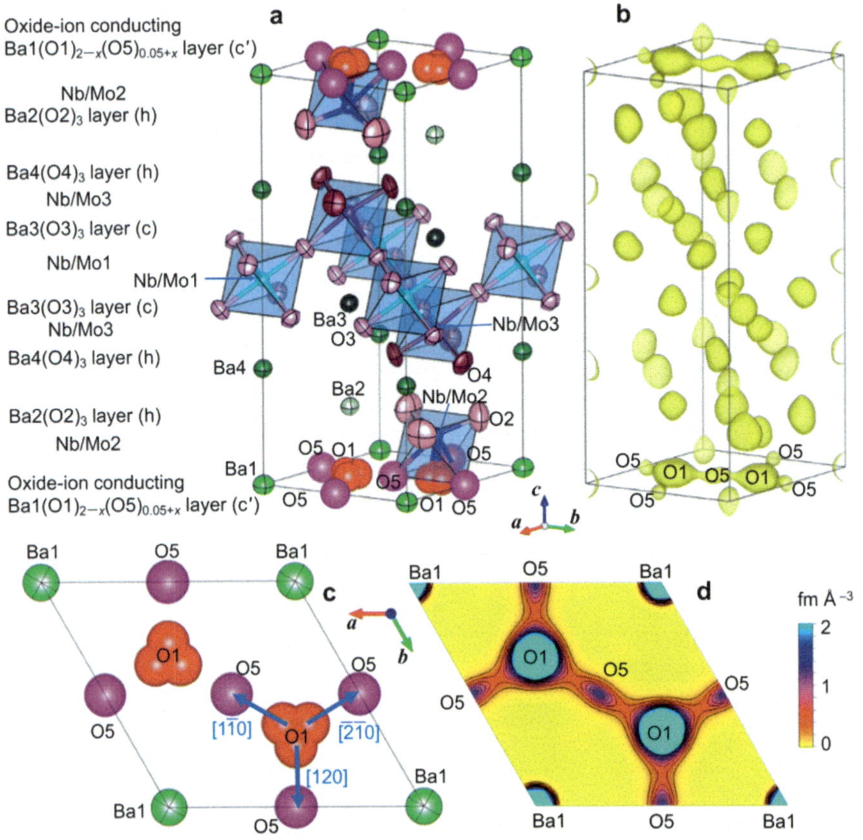


**Fig. 1.8** Experimental evidence of the interstitial oxygen O5 and the O1–O5 oxide-ion interstitialcy diffusion of $Ba_7Nb_{3.9}Mo_{1.1}O_{20.05}$ at a high temperature of 800 °C. **a** Refined crystal structure and **b** corresponding yellow isosurface of maximum-entropy method neutron scattering length densities (MEM NSLDs) at 0.36 fm Å$^{-3}$ of $Ba_7Nb_{3.9}Mo_{1.1}O_{20.05}$ at 800 °C. Refined crystal structure **c** and corresponding MEM NSLD distribution **d** on the *ab* plane at $z = 0$ of $Ba_7Nb_{3.9}Mo_{1.1}O_{20.05}$ at 800 °C. In **d**, the contour lines from 0 to 2 fm Å$^{-3}$ by the step of 0.2 fm Å$^{-3}$. Thermal ellipsoids in panels (a) and (c) are drawn at the 90% probability level. Arrows in **c** denote the directions of oxide-ion O1-to-O5 migration [63]. Copyright© 2021 Springer Nature

proton conduction, a portion of $Nb^{5+}$ was partially substituted with $Cr^{6+}$, which is hexavalent and has never been reported to exhibit proton conductivity.

$Ba_7Nb_{3.9}Mo_{1.1}O_{20.05}$, in which some of the $Nb^{5+}$ in $Ba_7Nb_4MoO_{20}$ is substituted with $Mo^{6+}$, exhibits a high conductivity of $5.8 \times 10^{-4}$ S cm$^{-1}$ at 310 °C. In addition, it is a highly stable material in which oxide-ion conduction dominates in the oxygen partial pressure range of 1–2 × 10$^{-26}$ atm. [63] However, the Nb/Mo ratio had not been optimized for oxide-ion conductivity, and the only mechanism known for oxide-ion conduction was the interstitialcy mechanism.

In this work, I optimized the composition and discovered that $Ba_7Nb_{3.8}Mo_{1.2}O_{20.1}$ exhibits the highest oxide-ion conductivity among $Ba_7Nb_4MoO_{20}$-based materials. By carefully examining snapshots by AIMD simulations, we dynamically and locally tracked the changes in polyhedrons due to Nb/Mo and oxygen near the c′ layer and considered the mechanism of oxide-ion migration. Furthermore, by carefully examining snapshots from the AIMD simulation, I traced the local and dynamic changes in the polyhedrons due to Nb/Mo and oxygen near the c′ layer and considered the mechanism of oxide-ion migration.

In addition, little consideration had been given to the factors that explain the high oxide ion conductivity at low temperatures, which is a characteristic of $Ba_7Nb_4MoO_{20}$-based materials, and it was discovered that the distance between Ba and Nb/Mo that sandwich the c' layer, where oxide ions diffuse, is the key. Furthermore, although $Ba_7Nb_4MoO_{20}$-based materials are dual ion conductors that also exhibit proton conductivity, the proton movement has only been considered in the c′ layer [64]. Therefore, by AIMD simulations of $(Ba_7Nb_{3.8}Mo_{1.2}O_{20.1}\cdot 0.1\ H_2O)_9$, the proton migration pathway was considered not only in the c′ layer but also in the close-packed layer.

## 1.10 The Purpose of This Work

Hexagonal perovskite-related oxides $Ba_7Nb_4MoO_{20}$-based materials with the intrinsically oxygen-deficient layer are dual ion conductors because they exhibit high oxide-ion conductivity and proton conductivity. However, the composition has not been optimized, and the ion migration mechanism has remained largely unknown.

In Chap. 2, I first attempted element substitution to simultaneously improve oxide-ion conductivity and suppress proton conductivity in order to discover a new material that only exhibits high oxide-ion conductivity. By substituting part of the $Nb^{5+}$ in $Ba_7Nb_4MoO_{20}$ with $Cr^{6+}$, I discovered $Ba_7Nb_{3.8}Cr_{0.2}MoO_{20.1}$. It was shown that $Ba_7Nb_{3.8}Cr_{0.2}MoO_{20.1}$ has improved oxide ion conductivity and suppressed proton conduction compared to the parent material $Ba_7Nb_4MoO_{20}$. Furthermore, to investigate the site of $Cr^{6+}$, the preliminary Rietveld analyses were performed to compare the reliability factor values where $Cr^{6+}$ is placed at the Nb/Mo1 site, Nb/Mo2 site, and Nb/Mo3 site. MEM analysis shows that oxide ions diffuse by interstitialcy mechanism.

In Chap. 3, by partially substituting the $Nb^{5+}$ in $Ba_7Nb_4MoO_{20}$ with $Mo^{6+}$, I discovered $Ba_7Nb_{3.8}Mo_{1.2}O_{20.1}$, which exhibits the highest oxide-ion conductivity and dual-ion conductivity among $Ba_7Nb_4MoO_{20}$-based materials. $Ba_7Nb_{3.8}Mo_{1.2}O_{20.1}$ was found to be highly stable even under reducing atmosphere, and ion conduction was dominant. The reason why $Ba_7Nb_{3.8}Mo_{1.2}O_{20.1}$ exhibits high oxide-ion conductivity was clarified by investigating the amount of interstitial oxygen and the minimum NSLD in oxide-ion diffusion path by Rietveld and MEM analyses. Furthermore, I investigated the oxide-ion migration mechanism from the viewpoint of local and dynamic structure in $Ba_7Nb_{3.8}Mo_{1.2}O_{20.1}$ by AIMD simulation, focusing on the coordination number of polyhedra. I also showed which layer the protons fast diffuse at high speed by the AIMD simulation of $(Ba_7Nb_{3.8}Mo_{1.2}O_{20.1}\cdot 0.1\ H_2O)_9$. The simulation results were supported by Rietveld and MEM analyses. In addition, it was shown that protons diffuse more easily in $(Ba_7Nb_{3.8}Mo_{1.2}O_{20.1}\cdot 0.1\ H_2O)_9$ than in the parent material $(Ba_7Nb_4MoO_{20}\cdot 0.1\ H_2O)_9$.

## References


1. https://www.meti.go.jp/policy/chemical_management/chemical_wondertown/labo/page02.html. Accessed 6 Dec 2023
2. Staffell I, Scamman D, Abad AV, Balcombe P, Dodds PE, Ekins P, Shah N, Ward KR (2019) The role of hydrogen and fuel cells in the global energy system. Energy Environ Sci 12:463–491
3. Coduri M, Karlsson M, Malavasi L (2022) Structure-property correlation in oxide-ion and proton conductors for clean energy applications: recent experimental and computational advancements. J Mater Chem A 10:5082–5110
4. Tubandt C, Lorenz EZ (1914) The molecular condition and electrical conductivity of crystallized salts. Z Phys Chem 87:513–542
5. Li Y, Song S, Kim H, Nomoto K, Kim H, Sun X, Hori S, Suzuki K, Matsui N, Hirayama M, Mizoguchi T, Saito T, Kamiyama T, Kanno R (2023) A lithium superionic conductor for millimeter-thick battery electrode. Science 381(6653):50–53
6. Ishihara T, Matsuda H, Takita Y (1994) Doped $LaGaO_3$ perovskite type oxide as a new oxide ionic conductor. J Am Chem Soc 116(9):3801–3803
7. Fop S, McCombie KS, Wildman EJ, Skakle JMS, Mclaughlin AC (2019) Hexagonal perovskite derivatives: a new direction in the design of oxide ion conducting materials. Chem Commun 55(15):2127–2137
8. Yaguchi H, Fujii K, Tsuchiya Y, Ogino H, Tsujimoto Y, Yashima M (2022) Ruddlesden-Popper oxychlorides $Ba_3Y_2O_5Cl_2$, $Sr_3Sc_2O_5Cl_2$, and $Sr_2ScO_3Cl$: first examples of oxide-ion-conducting oxychlorides. ACS Appl Energy Mater 5(1):295–304
9. Matsui M, Fujii K, Shiraiwa M, Yashima M (2022) Ge-containing oxide-ion conductors with $CaEu_2Ge_3O_{10}$-type structure discovered by the bond-valence method and experiments. Inorg Chem 61(31):12327–12336
10. Zhang W, Yashima M (2023) Recent developments in oxide ion conductors: focusing on Dion-Jacobson phases. Chem Commun 59:134–152
11. Yaguchi H, Morikawa D, Saito T, Tsuda K, Yashima M, Yaguchi H, Yashima M, Morikawa D, Saito T, Tsuda K (2023) High oxide-ion conductivity through the interstitial oxygen site in sillén oxychlorides. Adv Funct Mater 33:2214082
12. Kreuer KD (2003) Proton-conducting oxides. Annu Rev Mater 33(1):333–359
13. Fop S (2021) Solid oxide proton conductors beyond perovskites. J Mater Chem A 9(35):18836–18856
14. Liu J, Chang D, Whitfield P, Janssen Y, Yu X, Zhou Y, Bai J, Ko J, Nam KW, Wu L, Zhu Y, Feygenson M, Amatucci G, Van Der Ven A, Yang XQ, Khalifah P (2014) Ionic conduction in cubic $Na_3TiP_3O_9N$, a secondary Na-ion battery cathode with extremely low volume change. Chem Mater 26(10):3295–3305
15. Zhu Z, Chu IH, Deng Z, Ong SP (2015) Role of $Na^+$ interstitials and dopants in enhancing the $Na^+$ conductivity of the cubic $Na_3PS_4$ superionic conductor. Chem Mater 27(24):8318–8325
16. Balapanov MK, Gafurov IG, Mukhamed'yanov UK, Yakshibaev RA, Ishembetov RK (2004) Ionic conductivity and chemical diffusion in superionic $Li_xCu_{2-x}S$ $(0 \leq x \leq 0.25)$. Phys Status solidi B 241(1):114–119
17. Agrawal RC, Gupta RK, Kumar R, Kumar A (1994) Ionic transport in the (Agl:AgCl) mixed-system. J Mater Sci 29(14):3673–3677

18. Bobe JM, Réau JM, Senegas J, Poulain M (1995) $F^-$ ion conductivity and diffusion properties in $ZrF_4$-based fluoride glasses with various NaF concentrations ($0 \leq \chi_{NaF} \leq 0.45$). Solid State Ion 82(1–2):39–52
19. Jin W, Li Z, Wang Z, Fu YQ (2016) Mg ion dynamics in anode materials of Sn and Bi for Mg-Ion batteries. Mater Chem Phys 182:167–172
20. Zhan Y, Zhang W, Lei B, Liu H, Li W (2020) Recent development of Mg ion solid electrolyte. Front Chem 8:524782
21. Huang K, Feng M, Goodenough JB (1998) Synthesis and electrical properties of dense $Ce_{0.9}Gd_{0.1}O_{1.95}$ ceramics. J Am Ceram Soc 81(2):357–362
22. Yashima M, Ishimura D (2003) Crystal structure and disorder of the fast oxide-ion conductor cubic $Bi_2O_3$. Chem Phys Lett 378(3–4):395–399
23. Kwon OH, Choi GM (2006) Electrical conductivity of thick film YSZ. Solid State Ion 177(35–36):3057–3062
24. Jung DW, Duncan KL, Wachsman ED (2010) Effect of total dopant concentration and dopant ratio on conductivity of $(DyO_{1.5})_x$-$(WO_3)_y$-$(BiO_{1.5})_{1-x-y}$. Acta Mater 58(2):355–363
25. Ishihara T, Kilner JA, Honda M, Takita Y (1997) Oxygen surface exchange and diffusion in the new perovskite oxide ion conductor $LaGaO_3$. J Am Chem Soc 119(11):2747–2748
26. Huang K, Tichy RS, Goodenough JB (1998) Superior perovskite oxide-ion conductor; strontium- and magnesium-doped $LaGaO_3$: I, Phase relationships and electrical properties. J Am Ceram Soc 81(10):2565–2575
27. Ishihara T, Furutani H (1999) Improved oxide ion conductivity in $La_{0.8}Sr_{0.2}Ga_{0.8}Mg_{0.2}O_3$ by doping Co. Chem Mater 11(8):2081–2088
28. Ali R, Yashima M (2005) Space Group and Crystal Structure of the Perovskite $CaTiO_3$ from 296 to 1720 K. J Solid State Chem 178(9):2867–2872
29. Yashima M, Ali R (2009) Structural phase transition and octahedral tilting in the calcium titanate perovskite $CaTiO_3$. Solid State Ion 180(2–3):120–126
30. Zhang W, Fujii K, Ishiyama T, Kandabashi H, Yashima M (2020) Dion–Jacobson-type oxide-ion conductor $CsLa_2Ti_2NbO_{10-\delta}$ without phase transitions. J Mater Chem A 8(47):25085–25093
31. Zhang W, Fujii K, Niwa E, Hagihala M, Kamiyama T, Yashima M (2020) Oxide-ion conduction in the Dion-Jacobson phase $CsBi_2Ti_2NbO_{10-\delta}$. Nat Commun 11(1):1224
32. Skinner S, Kilner J (2009) Oxygen diffusion and surface exchange in $La_{2-x}Sr_xNiO_{4+\delta}$. Solid State Ion 125(1–4):709–712
33. Yashima M, Sirikanda N, Ishihara T (2010) Crystal structure, diffusion path, and oxygen permeability of a $Pr_2NiO_4$-based mixed conductor $(Pr_{0.9}La_{0.1})_2(Ni_{0.74}Cu_{0.21}Ga_{0.05})O_{4+\delta}$. J Am Chem Soc 132(7):2385–2392
34. Yashima M, Yamada H, Nuansaeng S, Ishihara T (2012) Role of $Ga^{3+}$ and $Cu^{2+}$ in the high interstitial oxide-ion diffusivity of $Pr_2NiO_4$-based oxides: design concept of interstitial ion conductors through the higher-valence $d^{10}$ dopant and Jahn-Teller effect. Chem Mater 24(21):4100–4113
35. Nakayama S, Sakamoto M (1998) Electrical properties of new type high oxide ionic conductor $RE_{10}Si_6O_{27}$ (RE = La, Pr, Nd, Sm, Gd, Dy). J Eur Ceram Soc 18(10):1413–1418
36. Ali R, Yashima M, Matsushita Y, Yoshioka H, Ohoyama K, Izumi F (2008) Diffusion path of oxide ions in an Apatite-type ionic conductor $La_{9.69}(Si_{5.70}Mg_{0.30})O_{26.24}$. Chem Mater 20(16):5203–5208
37. Jones A, Slater PR, Islam MS (2008) Local defect structures and ion transport mechanisms in the oxygen-excess apatite $La_{9.67}(SiO_4)_6O_{2.5}$. Chem Mater 20(15):5055–5060
38. Fujii K, Yashima M, Hibino K, Shiraiwa M, Fukuda K, Nakayama S, Ishizawa N, Hanashima T, Ohhara T (2018) High oxide-ion conductivity in Si-deficient $La_{9.565}(Si_{5.826}\square_{0.174})O_{26}$ apatite without interstitial oxygens due to the overbonded channel oxygens. J Mater Chem A 6(23):10835–10846
39. Kuang X, Green MA, Niu H, Zajdel P, Dickinson C, Claridge JB, Jantsky L, Rosseinsky MJ (2008) Interstitial oxide ion conductivity in the layered tetrahedral network melilite structure. Nat Mater 7(6):498–504

40. Li M-R, Kuang X, Chong SY, Xu Z, Thomas CI, Niu H, Claridge JB, Rosseinsky MJ (2010) Interstitial oxide ion order and conductivity in $La_{1.64}Ca_{0.36}Ga_3O_{7.32}$ Melilite. Angew Chem 122(13):2412–2416
41. Xu J, Wang J, Rakhmatullin A, Ory S, Fernández-Carrión AJ, Yi H, Kuang X, Allix M (2019) Interstitial oxide ion migration mechanism in aluminate Melilite $La_{1+x}Ca_{1-x}Al_3O_{7+0.5x}$ ceramics synthesized by glass crystallization. ACS Appl Energy Mater 2(4):2878–2888
42. Zhou L, Xu J, Allix M, Kuang X (2020) Development of Melilite-type oxide ion conductors. Chem Rec 20(10):1117–1128
43. Goutenoire F, Isnard O, Retoux R, Lacorre P (2000) Crystal structure of $La_2Mo_2O_9$, a new fast oxide-ion conductor. Chem Mater 12(9):2575–2580
44. Goutenoire F, Isnard O, Suard E, Bohnke O, Laligant Y, Retoux R, Lacorre P (2001) Structural and transport characteristics of the LAMOX family of fast oxide-ion conductors, based on lanthanum molybdenum oxide $La_2Mo_2O_9$. J Mater Chem 11(1):119–124
45. Georges S, Skinner SJ, Lacorre P, Steil MC (2004) Oxide ion diffusion in optimised LAMOX materials. Dalton Trans 7(19):3101–3105
46. Esaka T, Mina-ai T, Iwahara H (1992) Oxide ion conduction in the solid solution based on the Scheelite-type oxide $PbWO_4$. Solid State Ion 57(3–4):319–325
47. Esaka T (2000) Ionic conduction in substituted Scheelite-type oxides. Solid State Ion 136–137:1–9
48. Packer RJ, Skinner SJ, Yaremchenko AA, Tsipis EV, Kharton VV, Patrakeev MV, Bakhteeva YA (2006) Lanthanum substituted $CeNbO_{4+\delta}$ Scheelites: mixed conductivity and structure at elevated temperatures. J Mater Chem 16(34):3503–3511
49. Toyoura K, Sakakibara Y, Yokoi T, Nakamura A, Matsunaga K (2018) Oxide-Ion conduction via interstitials in Scheelite-type $LaNbO_4$: a first-principles study. J Mater Chem A 6(25):12004–12011
50. Yang X, Fernández-Carrión AJ, Wang J, Porcher F, Fayon F, Allix M, Kuang X (2018) Cooperative mechanisms of oxygen vacancy stabilization and migration in the isolated tetrahedral anion Scheelite structure. Nat Commun 9(1):4484
51. Yang X, Zeng X, Ming X, Yang L, Fernández-Carrión AJ, Deng S, He L, Kuang X (2022) Oxide-ion conductivity optimization in $BiVO_4$ Scheelite by an acceptor doping strategy. Inorg Chem Front 9(11):2644–2658
52. Uno W, Fujii K, Niwa E, Torii S, Miao P, Kamiyama T, Yashima M (2018) Experimental visualization of oxide-ion diffusion paths in pyrochlore-type $Yb_2Ti_2O_7$. J Ceram Soc Jpn 126(5):341–345
53. Fujii K, Esaki Y, Omoto K, Yashima M, Hoshikawa A, Ishigaki T, Hester JR (2014) New perovskite-related structure family of oxide-ion conducting materials $NdBaInO_4$. Chem Mater 26(8):2488–2491
54. Fujii K, Shiraiwa M, Esaki Y, Yashima M, Kim SJ, Lee S (2015) Improved oxide-ion conductivity of $NdBaInO_4$ by Sr doping. J Mater Chem A 3(22):11985–11990
55. Shiraiwa M, Fujii K, Esaki Y, Kim SJ, Lee S, Yashima M (2017) Crystal structure and oxide-ion conductivity of $Ba_{1+x}Nd_{1-x}InO_{4-x/2}$. J Electrochem Soc 164(13):1392–1399
56. Kuang X, Bridges C, Allix M, Claridge JB, Hughes H, Rosseinsky MJ (2006) Internal barrier layer capacitance effect in hexagonal perovskite $Ba_4YMn_3O_{11.5}$. Chem Mater 18(21):5130–5136
57. Kuang X, Allix M, Ibberson RM, Claridge JB, Niu H, Rosseinsky MJ (2007) Oxygen vacancy ordering phenomena in the mixed-conducting hexagonal perovskite $Ba_7Y_2Mn_3Ti_2O_{20}$. Chem Mater 19(11):2884–2893
58. Fop S, Skakle JMS, McLaughlin AC, Connor PA, Irvine JTS, Smith RI, Wildman EJ (2016) Oxide ion conductivity in the hexagonal perovskite derivative $Ba_3MoNbO_{8.5}$. J Am Chem Soc 138(51):16764–16769
59. Chambers MS, McCombie KS, Auckett JE, McLaughlin AC, Irvine JTS, Chater PA, Evans JSO, Evans IR (2019) Hexagonal perovskite related oxide ion conductor $Ba_3NbMoO_{8.5}$: phase transition, temperature evolution of the local structure and properties. J Mater Chem A 7(44):25503–25510

60. Yashima M, Tsujiguchi T, Fujii K, Niwa E, Nishioka S, Hester JR, Maeda K (2019) Direct evidence for two-dimensional oxide-ion diffusion in the hexagonal perovskite-related oxide $Ba_3MoNbO_{8.5-\delta}$. J Mater Chem A 7(23):13910–13916
61. Gilane A, Fop S, Sher F, Smith RI, McLaughlin AC (2020) The relationship between oxide-ion conductivity and cation vacancy order in the hybrid hexagonal perovskite $Ba_3VWO_{8.5}$. J Mater Chem A 8(32):16506–16514
62. Fop S, McCombie KS, Wildman EJ, Skakle JMS, Irvine JTS, Connor PA, Savaniu C, Ritter C, Mclaughlin AC (2020) High oxide ion and proton conductivity in a disordered hexagonal perovskite. Nat Mater 19(7):752–757
63. Yashima M, Tsujiguchi T, Sakuda Y, Yasui Y, Zhou Y, Fujii K, Torii S, Kamiyama T, Skinner SJ (2021) High oxide-ion conductivity through the interstitial oxygen site in $Ba_7Nb_4MoO_{20}$-based hexagonal perovskite related oxides. Nat Commun 12:556
64. Fop S, Dawson JA, Fortes AD, Ritter C, McLaughlin AC (2021) Hydration and ionic conduction mechanisms of hexagonal perovskite derivatives. Chem Mater 33(12):4651–4660
65. Murakami T, Shibata T, Yasui Y, Fujii K, Hester JR, Yashima M (2021) High oxide-ion conductivity in a hexagonal perovskite-related oxide $Ba_7Ta_{3.7}Mo_{1.3}O_{20.15}$ with cation site preference and interstitial oxide ions. Small 18(10):2106785
66. Nakayama S (2022) Characteristics of electrolyte-supported SOFC single cells using hexagonal perovskite $Ba_3MoNbO_{8.5}$ and $Ba_7Nb_4MoO_{20}$ ceramics. Results Mater 16:100318
67. Tansho M, Goto A, Ohki S, Mogami Y, Sakuda Y, Yasui Y, Murakami T, Fujii K, Iijima T, Yashima M (2022) Different local structures of Mo and Nb polyhedra in the oxide-ion-conducting hexagonal perovskite-related oxide $Ba_3MoNbO_{8.5}$ revealed by $^{95}Mo$ and $^{93}Nb$ NMR measurements. J Phys Chem C 126(31):13284–13290
68. Suzuki Y, Murakami T, Fujii K, Hester JR, Yasui Y, Yashima M (2022) Simultaneous reduction of proton conductivity and enhancement of oxide-ion conductivity by aliovalent doping in $Ba_7Nb_4MoO_{20}$. Inorg Chem 61(19):7537–7545
69. Yasui Y, Tsujiguchi T, Sakuda Y, Hester JR, Yashima M (2022) Oxide-ion occupational disorder, diffusion path, and conductivity in hexagonal perovskite derivatives $Ba_3WNbO_{8.5}$ and $Ba_3MoNbO_{8.5}$. J Phys Chem C 126(5):2383–2393
70. Yasui Y, Tansho M, Fujii K, Sakuda Y, Goto A, Ohki S, Mogami Y, Iijima T, Kobayashi S, Kawaguchi S, Osaka K, Ikeda K, Otomo T, Yashima M (2023) Hidden chemical order in disordered $Ba_7Nb_4MoO_{20}$ revealed by resonant X-ray diffraction and solid-state NMR. Nat Commun 14:2337
71. Kikuchi Y, Yasui Y, Hester JR, Yashima M (2023) Improved oxide ion conductivity of hexagonal perovskite-related oxides $Ba_3W_{1+x}V_{1-x}O_{8.5+x/2}$. Inorganics 11(6):238
72. Li M, Pietrowski MJ, De Souza RA, Zhang H, Reaney IM, Cook SN, Kilner JA, Sinclair DC (2014) A family of oxide ion conductors based on the ferroelectric perovskite $Na_{0.5}Bi_{0.5}TiO_3$. Nat Mater 13(1):31–35
73. Lacorre P, Goutenoire F, Bohnke O, Retoux R, Lallgant Y (2000) Designing fast oxide-ion conductors based on $La_2Mo_2O_9$. Nature 404(6780):856–858
74. López CA, Pedregosa JC, Lamas DG, Alonso JA (2014) The strongly defective double perovskite $Sr_{11}Mo_4O_{23}$: crystal structure in relation to ionic conductivity. J Appl Cryst 47(4):1395–1401
75. Goodenough JB, Ruiz-Diaz JE, Zhen YS (1990) Oxide-ion conduction in $Ba_2In_2O_5$ and $Ba_3In_2MO_8$ (M = Ce, Hf, or Zr). Solid State Ion 44(1–2):21–31
76. León-Reina L, Losilla ER, Martínez-Lara M, Martín-Sedeño MC, Bruque S, Núñez P, Sheptyakov DV, Aranda MAG (2005) High oxide ion conductivity in Al-doped germanium oxyapatite. Chem Mater 17(3):596–600
77. Iwahara H (1995) Technological challenges in the application of proton conducting ceramics. Solid State Ion 77(C):289–298
78. Iwahara H, Asakura Y, Katahira K, Tanaka M (2004) Prospect of hydrogen technology using proton-conducting ceramics. Solid State Ion 168(3–4):299–310
79. Zhou Y, Shiraiwa M, Nagao M, Fujii K, Tanaka I, Yashima M, Baque L, Basbus JF, Mogni LV, Skinner SJ (2021) Protonic conduction in the $BaNdInO_4$ structure achieved by acceptor doping. Chem Mater 33(6):2139–2146

80. Shiraiwa M, Kido T, Fujii K, Yashima M (2021) High-temperature proton conductors based on the (110) layered perovskite $BaNdScO_4$. J Mater Chem A 6:4883–5230
81. Amsif M, Marrero-López D, Magrasó A, Peña-Martínez J, Ruiz-Morales JC, Núñez P (2009) Synthesis and characterisation of $BaCeO_3$-based proton conductors obtained from freeze-dried precursors. J Eur Ceram Soc 29(1):131–138
82. Zhong Z, Li Z, Li J, Guo X, Hu Q, Feng Y, Sun H (2022) A facile method to synthesize $BaZr_{0.1}Ce_{0.7}Y_{0.1}Yb_{0.1}O_{3-\delta}$ (BZCYYb) nanopowders for the application on highly conductive proton-conducting electrolytes. Int J Hydrogen Energy 47(94):40054–40066
83. Saito K, Yashima M (2023) High proton conductivity within the 'Norby Gap' by stabilizing a perovskite with disordered intrinsic oxygen vacancies. Nat Commun 14(1):7466
84. Brand AD, Antunes I, Frade JR, Torre J, Kharton VV, Fagg DP (2010) Enhanced low-temperature proton conduction in $Sr_{0.02}La_{0.98}NbO_{4-\delta}$ by Scheelite phase retention. Chem Mater 22:6673
85. Huse M, Norby T, Haugsrud R (2011) Proton conductivity in acceptor-doped $LaVO_4$. J Electrochem Soc 158(8):B857
86. Schober T, Friedrich J, Krug F (1997) Phase transition in the oxygen and proton conductor $Ba_2In_2O_5$ in humid atmospheres below 300 °C. Solid State Ion 99(1–2):9–13
87. Kendrick E, Kendrick J, Knight KS, Islam MS, Slater PR (2007) Cooperative mechanisms of fast-ion conduction in gallium-based oxides with tetrahedral moieties. Nat Mater 6(11):871–875
88. Yashima M, Kubo N, Omoto K, Fujimori H, Fujii K, Ohoyama K (2014) Diffusion path and conduction mechanism of protons in hydroxyapatite. J Phys Chem C 118(10):5180–5187
89. Ling CD, Avdeev M, Kutten R, Kharton VV, Yaremchenko AA, Fialkova S, Sharma N, Macquart RB, Hoelzel M, Gutmann M (2009) Structures, phase transitions, hydration, and ionic conductivity of $Ba_4Nb_2O_9$. Chem Mater 21(16):3853–3864
90. Ling CD, Avdeev M, Kharton VV, Yaremchenko AA, Macquart RB, Hoelzel M (2010) Structures, phase transitions, hydration, and ionic conductivity of $Ba_4Ta_2O_9$. Chem Mater 22(2):532–540
91. Wind J, Mole RA, Yu D, Avdeev M, Ling CD (2018) Hydration mechanisms and proton conduction in the mixed ionic-electronic conductors $Ba_4Nb_2O_9$ and $Ba_4Ta_2O_9$. Chem Mater 30(15):4949–4958
92. Murakami T, Hester JR, Yashima M (2020) High proton conductivity in $Ba_5Er_2Al_2ZrO_{13}$, a hexagonal perovskite-related oxide with intrinsically oxygen-deficient layers. J Am Chem Soc 142(27):11653–11657
93. Morikawa R, Murakami T, Fujii K, Avdeev M, Ikeda Y, Nambu Y, Yashima M (2023) High proton conduction in $Ba_2LuAlO_5$ with highly oxygen-deficient layers. Commun Mater 4(1):42
94. Murakami T, Avdeev M, Morikawa R, Hester JR, Yashima M, Murakami T, Avdeev M, Morikawa R, Yashima M, Hester JR (2023) High proton conductivity in β-$Ba_2ScAlO_5$ enabled by octahedral and intrinsically oxygen-deficient layers. Adv Funct Mater 33(7):2206777
95. Yamazaki Y, Hernandez-Sanchez R, Haile SM (2009) High total proton conductivity in large-grained yttrium-doped barium zirconate. Chem Mater 21(13):2755–2762
96. Cao J, Su C, Ji Y, Yang G, Shao Z (2021) Recent advances and perspectives of fluorite and perovskite-based dual-ion conducting solid oxide fuel cells. J Energy Chem 57:406–427
97. Zhou C, Sunarso J, Song Y, Dai J, Zhang J, Gu B, Zhou W, Shao Z (2019) New reduced-temperature ceramic fuel cells with dual-ion conducting electrolyte and triple-conducting double perovskite cathode†. J Mater Chem A 7:13265–13274
98. Yang L, Wang S, Blinn K, Liu M, Liu Z, Cheng Z, Liu M (2009) Enhanced sulphur and coking tolerance of a mixed ion conductor for SOFCs: $BaZr_{0.1}Ce_{0.7}Y_{0.2-x}Yb_xO_{3-\delta}$. Science 326(5949):126–129
99. Choi, S.; Kucharczyk, C. J.; Liang, Y.; Zhang, X.; Takeuchi, I.; Ji, H. Il; Haile, S. M. Exceptional Power Density and Stability at Intermediate Temperatures in Protonic Ceramic Fuel Cells. Nat. Energy. 2018, 3 (3), 202–210.
100. Seong A, Kim J, Jeong D, Sengodan S, Liu M, Choi S, Kim G (2021) Electrokinetic proton transport in triple ($H^+/O^{2-}/e^-$) conducting oxides as a key descriptor for highly efficient protonic ceramic fuel cells. Adv Sci 8(11):2004099

101. Fuller CA, Blom DA, Vogt T, Evans IR, Evans JSO (2022) Oxide ion and proton conductivity in a family of highly oxygen-deficient perovskite derivatives. J Am Chem Soc 144(1):615–624
102. Li S, Schönberger F, Slater P (2003) $La_{1-x}Ba_{1+x}GaO_{4-x/2}$: a novel high temperature proton conductor. Chem Commun 21:2694–2695
103. Noirault S, Célérier S, Joubert O, Caldes MT, Piffard Y (2007) Incorporation of water and fast proton conduction in the inherently oxygen-deficient compound $La_{26}O_{27}\square(BO_3)_8$. Adv Mater 19(6):867–870
104. López-Vergara A, Porras-Vázquez JM, Infantes-Molina A, Canales-Vázquez J, Cabeza A, Losilla ER, Marrero-López D (2017) Effect of preparation conditions on the polymorphism and transport properties of $La_{6-x}MoO_{12-\delta}$ $(0 \leq x \leq 0.8)$. Chem Mater 29(16):6966–6975
105. Jacucci G, Rahman A (1978) Diffusion of F-ions in $CaF_2$. J Chem Phys 69(9):4117–4125
106. Yashima M, Itoh M, Inaguma Y, Morii Y (2005) Crystal structure and diffusion path in the fast lithium-ion conductor $La_{0.62}Li_{0.16}TiO_3$. J Am Chem Soc 127(10):3491–3495
107. Kendrick E, Islam MS, Slater PR (2008) Atomic-scale mechanistic features of oxide ion conduction in apatite-type germanates. Chem Commun 6:715–717
108. Yashima M (2008) Crystal structures, structural disorders and diffusion paths of ionic conductors from diffraction experiments. Solid State Ion 179(21–26):797–803
109. Ali R, Yashima M, Matsushita Y, Yoshioka H, Izumi F (2009) Crystal structure and electron density in the apatite-type ionic conductor $La_{9.71}(Si_{5.81}Mg_{0.18})O_{26.37}$. J Solid State Chem 182(10):2846–2851
110. Chen YC, Yashima M, Ohta T, Ohoyama K, Yamamoto S (2012) Crystal structure, oxygen deficiency, and oxygen diffusion path of perovskite-type lanthanum cobaltites $La_{0.4}Ba_{0.6}CoO_{3-\delta}$ and $La_{0.6}Sr_{0.4}CoO_{3-\delta}$. J Phys Chem C 116(8):5246–5254
111. Shimojo F, Aniya M (2013) Diffusion mechanism of Ag ions in superionic conductor $Ag_2Se$ from Ab initio molecular-dynamics simulations. 74(4):1224–1230
112. Jalem R, Yamamoto Y, Shiiba H, Nakayama M, Munakata H, Kasuga T, Kanamura K (2013) Concerted migration mechanism in the Li ion dynamics of garnet-type $Li_7La_3Zr_2O_{12}$. Chem Mater 25(3):425–430
113. Chen Y-C, Yashima M, Peñ A-Martínez J, Kilner JA (2013) Experimental visualization of the diffusional pathway of oxide ions in a layered perovskite-type cobaltite $PrBaCo_2O_{5+\delta}$. 25(13):2638–2641
114. Yashima M (2015) Invited review: some recent developments in the atomic-scale characterization of structural and transport properties of ceria-based catalysts and ionic conductors. Catal Today 253:3–19
115. Fujii K, Yashima M (2018) Discovery and development of $BaNdInO_4$—a brief review. J Ceram Soc Jpn 126(10):852–859
116. Zhang Z, Zou Z, Kaup K, Xiao R, Shi S, Avdeev M, Hu Y, Wang D, He B, Li H, Huang X, Nazar LF, Chen L (2019) Correlated migration invokes higher $Na^+$-ion conductivity in NaSICON-type solid electrolytes. Adv Energy Mater 9(42)
117. Mukai K, Yashima M, Hibino K, Terai T (2019) Experimental visualization of interstitialcy diffusion of Li ion in $\beta$-$Li_2TiO_3$. ACS Appl Energy Mater 2(8):5481–5489
118. Gao S, Broux T, Fujii S, Tassel C, Yamamoto K, Xiao Y, Oikawa I, Takamura H, Ubukata H, Watanabe Y, Fujii K, Yashima M, Kuwabara A, Uchimoto Y, Kageyama H (2021) Hydride-based antiperovskites with soft anionic sublattices as fast alkali ionic conductors. Nat Commun 12:151
119. Nazrul Islam SMK, Mayank P, Ouyang Y, Chen J, Sagotra AK, Li M, Cortie MB, Mole R, Cazorla C, Yu D, Wang X, Robinson RA, Cortie DL (2021) Copper diffusion rates and hopping pathways in superionic $Cu_2Se$. Acta Mater 215:117026
120. Fop S, Dawson JA, Tawse DN, Skellern MG, Skakle JMS, McLaughlin AC (2022) Proton and oxide ion conductivity in Palmierite oxides. Chem Mater 34(18):8190–8197
121. Famprikis T, Canepa P, Dawson JA, Saiful Islam M, Masquelier C (2019) Fundamentals of inorganic solid-state electrolytes for batteries. Nat Mater 18:1278–1291
122. Wei H, Hao J, Yang J, Lv Y, Shen F, Zhang W, Chen J, He L, Liu L, Kuang X, Xu J (2023) Thermally induced oxygen vacancies and high oxide ion conduction in $K_2ZnV_2O_7$ with a melilite-related structure. ACS Mater Au 3(5):492–500

123. Yashima M, Enoki M, Wakita T, Ali R, Matsushita Y, Izumi F, Ishihara T (2008) Structural disorder and diffusional pathway of oxide ions in a doped $Pr_2NiO_4$-based mixed conductor. J Am Chem Soc 130(9):2762–2763
124. Chroneos A, Parfitt D, Kilner JA, Grimes RW (2009) Anisotropic oxygen diffusion in tetragonal $La_2NiO_{4+\delta}$: molecular dynamics calculations. J Mater Chem 20(2):266–270
125. Li J, Pan F, Geng S, Lin C, Palatinus L, Allix M, Kuang X, Lin J, Sun J (2020) Modulated structure determination and ion transport mechanism of oxide-ion conductor $CeNbO_{4+\delta}$. Nat Commun 11:4751
126. Tilley RJD (2016) Perovskites: structure-property relationships. Wiley, Chichester, UK.
127. Yang X, Fernández-Carrión AJ, Kuang X (2023) Oxide ion-conducting materials containing tetrahedral moieties: structures and conduction mechanisms. Chem Rev

# Chapter 2
# Improved Oxide-Ion and Lower Proton Conduction of Hexagonal Perovskite-Related Oxides Based on $Ba_7Nb_4MoO_{20}$ by $Cr^{6+}$ Doping

**Abstract** Oxide-ion conductors based on hexagonal perovskite-related oxide $Ba_7Nb_4MoO_{20}$ have attracted much attention due to high oxide-ion and proton conductivities and potential applications in many electrochemical devices such as solid oxide fuel cells (SOFCs). Herein, we report simultaneous improvement of oxide-ion conductivity and suppression of proton conductivity by $Cr^{6+}$ doping in $Ba_7Nb_4MoO_{20}$. New materials $Ba_7Nb_{4-x}MoCr_xO_{20+x/2}$ ($x = 0.05, 0.1, 0.15, 0.2, 0.25, 0.3, 0.35, 0.4, 0.45, 0.5$) were synthesized. It was found that $Ba_7Nb_{3.8}Cr_{0.2}MoO_{20.1}$ exhibits high oxide-ion conductivity of $1.6 \times 10^{-3}$ S cm$^{-1}$ at 508 °C and $1.1 \times 10^{-2}$ S cm$^{-1}$ at 904 °C in static air and low proton transport number under wet conditions. $Ba_7Nb_{3.8}Cr_{0.2}MoO_{20.1}$ also shows wide electrolyte domain in the oxygen partial pressure $P(O_2)$ regions from 1 to $2.2 \times 10^{-27}$ atm (304 °C) and 1 to $1.5 \times 10^{-26}$ atm (604 °C), indicating extremely high chemical and electrical stability. The structure analyses have shown that $Ba_7Nb_{3.8}Cr_{0.2}MoO_{20.1}$ is a hexagonal perovskite-related oxide at 22 and 800 °C. Refined crystal structure of $Ba_7Nb_{3.8}Cr_{0.2}MoO_{20.1}$ has oxygen-deficient cubic (c′) close-packed $Ba(O1)_{2-y}(O5)_z$ layer where $y$ is the amount of oxygen vacancy at the tetrahedral O1 site and $z$ is the amount of interstitial octahedral oxygen atoms at an O5 site. The bond valence sum of Cr cation indicates the oxidation number of + 6: $Cr^{6+}$. The $Cr^{6+}$ cation is located at a crystallographic site near the c′ layer, which leads to the excess oxygen and high conductivity, and is likely to suppress the proton conduction. Maximum-entropy method (MEM) analyses have demonstrated that oxide ions two-dimensionally migrate through the lattice O1 and interstitial O5 sites in the c′ layer via the interstitialcy diffusion mechanism at 800 °C, which enables the high oxide-ion conduction. The $Cr^{6+}$ doping in various hexagonal perovskite-related oxides would be a new strategy for the simultaneous improvement of oxide-ion conductivity and suppression of proton conduction.



Y. Sakuda, *Discovery of $Ba_7Nb_4MoO_{20}$-Based Materials and the Mechanism of Ultrafast Ion Conduction via Dimers*, Springer Theses,
https://doi.org/10.1007/978-981-96-2552-9_2

## 2.1 Introduction

Oxide-ion and proton conductors have found a wide variety of applications such as sensors, catalysts, and solid oxide fuel cells (SOFCs) [1–29]. To develop better electrolytes in conventional SOFCs, oxide-ion conductors exhibiting higher conductivities and lower proton conductivities are required. Recently, oxide-ion conductors based on hexagonal perovskite-related oxide $Ba_7Nb_4MoO_{20}$ have attracted much attention due to high oxide-ion and proton conductivities [22–26]. The crystal structure of a hexagonal perovskite-related compound $A_mM_nX_p$ has hexagonal closed-packed $AX_3$ (h) layers and/or anion-deficient $AX_{3-\varepsilon}$ layers where $A$ is a larger cation, $M$ is a smaller cation, $X$ is an anion, $m$, $n$, and $p$ are positive numbers, and $\varepsilon$ is the amount of anion deficiency. The crystal structures of $Ba_7Nb_4MoO_{20}$ and its related materials have the stacking of h layers, cubic close-packed (c) layers and oxygen deficient cubic close-packed (c′) layers [22–26]. Oxide-ion and proton migrate in the c′ layer in $Ba_7Nb_4MoO_{20}$ and related materials [22–26]. $Ba_7Nb_4MoO_{20}$ exhibits both oxide-ion and proton conduction, which might reduce the performance of conventional SOFCs. Therefore, simultaneous improvement of oxide-ion conduction and suppression of proton conductivity are necessary for better electrolytes of SOFCs. The substitution of $Cr^{6+}$ for $Nb^{5+}$ ($Cr^{6+}$ doping) in $Ba_7Nb_4MoO_{20}$ makes the larger amount of excess $O^{2-}$ anions, larger amount of interstitial oxygen and smaller amount of oxygen vacancies in the c′ layer. Thus, the higher oxide-ion conductivity and lower proton conductivity are expected by the $Cr^{6+}$ doping into $Ba_7Nb_4MoO_{20}$.

The aims of this work is to (i) synthesize new hexagonal perovskite-related oxides $Ba_7Nb_{4-x}Cr_xMoO_{20+x/2}$ ($x$ = 0.05, 0.1, 0.15, 0.2, 0.25, 0.3, 0.35, 0.4, 0.45, 0.5), (ii) investigate the existing phase and lattice parameters of $Ba_7Nb_{4-x}Cr_xMoO_{20+x/2}$, (iii) examine the transport properties of $Ba_7Nb_{4-x}Cr_xMoO_{20+x/2}$, and (iv) analyze the crystal structure and neutron scattering length density distribution of $Ba_7Nb_{3.8}Cr_{0.2}MoO_{20.1-\delta}$ ($x = 0.2$) where $\delta$ is the amount of oxygen deficiency. The present work demonstrates simultaneous improvement of oxide-ion conductivity and suppression of proton conduction by Cr doping into the hexagonal perovskite related oxide $Ba_7Nb_4MoO_{20}$.

## 2.2 Methods

$Ba_7Nb_{4-x}Cr_xMoO_{20+x/2}$ ($x$ = 0, 0.05, 0.1, 0.15, 0.2, 0.25, 0.3, 0.35, 0.4, 0.45, 0.5) samples were synthesized by the solid-state-reaction method. Stoichiometric amounts of $BaCO_3$, $Nb_2O_5$, $MoO_3$, and $Cr_2O_3$ (> 99.9%, Kojundo Chemical Laboratory Co., Ltd.) were mixed and ground as ethanol slurries and dried powders using an agate mortar for 1 h, followed by calcination at 900 °C in air for 12 h. The obtained samples were ground again as ethanol slurries and dried powders with the agate mortar for 1 h, pressed uniaxially into pellets at about 150 MPa, and sintered in air at 1100 °C for 24 h. A few sintered pellets were crushed with a WC crusher and ground as ethanol slurries and dried powders with the agate mortar for 1 h, and then ball-milled using YSZ balls with a diameter of 5 mm for 20 min, 3 mm for 20 min, and 1 mm for 60 min (Fritsch, P7). The powders thus obtained were isostatically pressed into pellets at 150 MPa and subsequently sintered in air at 1200 °C for 12 h. Parts of the pellets were crushed with the WC crusher and ground into powders with the agate mortar to measure X-ray powder diffraction (XRD) data and X-ray photoelectron spectroscopy (XPS) spectra.

The phase purities of $Ba_7Nb_{4-x}Cr_xMoO_{20+x/2}$ ($x = 0, 0.05, 0.1, 0.15, 0.2, 0.25, 0.3, 0.35, 0.4, 0.45, 0.5$) were examined at room temperature (RT) by a laboratory-based X-ray diffractometer (Bruker AXS D8 Advance, Cu $K\alpha$ radiation). Lattice parameters were determined by Le Bail analyses using the computer program RIETAN-FP [29] and XRD data of $Ba_7Nb_{4-x}Cr_xMoO_{20+x/2}$ ($x = 0, 0.05, 0.1, 0.15, 0.2, 0.25, 0.3, 0.35, 0.4, 0.45, 0.5$) samples mixed with Si internal standard at RT. XPS spectra of $Ba_7Nb_{3.8}Cr_{0.2}MoO_{20.1}$ were measured by a JPS 9010 (JEOL Ltd.) X-ray photoelectron spectrometer. TG measurements of $Ba_7Nb_{3.8}Cr_{0.2}MoO_{20.1-\delta}$ were performed under dry air flow (100 mL $min^{-1}$) using a Bruker-AXS 2020SA from RT to 900 °C (3 cycles, heating and cooling rates of 2 °C $min^{-1}$).

The samples sintered at 1200 °C (relative density > 95%) were used for measurements of DC electrical conductivities $\sigma$. Pt paste and Pt wires were attached to the sintered pellets and heated in static air at 1000 °C for 1 h to remove organics from the Pt paste. Temperature dependencies of $\sigma$ in $Ba_7Nb_{4-x}Cr_xMoO_{20+x/2-\delta}$ ($x = 0, 0.05, 0.1, 0.15, 0.25, 0.35$, and $0.5$) were measured on heating by the direct current DC 4-terminal method in static air. The $\sigma$ data of $Ba_7Nb_{4-x}Cr_xMoO_{20+x/2-\delta}$ ($x = 0.2$) were collected on heating in static air, dry and wet air flow (100 mL $min^{-1}$, water vapor pressure $P(H_2O)$ in wet air flow was 0.021 atm) as a function of temperature. The $\sigma$ of the $Ba_7Nb_{3.8}Cr_{0.2}MoO_{20.1-\delta}$ pellet was measured at different oxygen partial pressures $P(O_2)$ at 304 and 604 °C. The $P(O_2)$ was controlled using a mixture of $O_2$, $N_2$, and 5% $H_2$ in $N_2$ gas, and monitored by an oxygen sensor placed at the outlet of the apparatus.

Neutron-diffraction data of $Ba_7Nb_{3.8}Cr_{0.2}MoO_{20.1}$ pellets sintered at 1100 °C were measured at 22 and 800 °C with the Echidna diffractometer [30] (wavelength: $\lambda = 1.622796(15)$ Å, step interval: 0.125° in $2\theta$). The neutron-diffraction data were analyzed by the Rietveld method using RIETAN-FP [31]. The neutron scattering length density distribution was investigated using the maximum-entropy method (MEM) with Dysnomia program [32]. The MEM calculations were performed with the unit cell divided into 60 × 60 × 168 pixels. Bond-valence-based energy (BVE) landscapes were calculated with SoftBV [33, 34] where the spatial resolution was set at approximately 0.1 Å. The crystal structures, neutron scattering length density distributions, and BVE landscapes were visualized using the VESTA program [35].

## 2.3 Results and Discussion

Figure 2.1a shows the XRD patterns of $Ba_7Nb_{4-x}MoCr_xO_{20+x/2}$ ($x = 0$, 0.05, 0.1, 0.15, 0.2, 0.25, 0.3, 0.35, 0.4, 0.45, 0.5) measured at RT. Most of the reflections are indexed based on the primitive hexagonal *hP* lattice, while small amounts of impurity phases such as $BaMoO_4$, $Ba_3MoNbO_{8.5}$, and/or $Ba_5Nb_4O_{15}$ are also observed. Therefore, the main phase in each composition $Ba_7Nb_{4-x}Cr_xMoO_{20+x/2}$ is identified to be 7H-polytype (crystal system: trigonal, space group: $P\overline{3}m1$) [22, 23, 26, 36]. The lattice volume of $Ba_7Nb_{4-x}Cr_xMoO_{20+x/2}$ decreases linearly with an increase of Cr content $x$ up to $x = 0.4$ (Fig. 2.1b; Table 2.1), indicating the formation of solid solutions $Ba_7Nb_{4-x}Cr_xMoO_{20+x/2}$ ($0.05 \leq x \leq 0.4$). The decrease of the lattice volume is ascribed to the smaller ionic radius of $Cr^{6+}$ (0.26 Å in fourfold coordination) in comparison with that of $Nb^{5+}$ (0.48 Å in fourfold coordination) [37]. For $x$ values higher than 0.4, the lattice volume remains almost constant, and the XRD peak intensities of impurity phases increase with $x$. The solubility limit in $Ba_7Nb_{4-x}Cr_xMoO_{20+x/2}$ is estimated to be $x = 0.394(14)$ from the extrapolation of a straight-line fit to the lattice volumes.

Figure 2.2a shows Arrhenius plots of the electrical conductivities $\sigma$ of $Ba_7Nb_{4-x}Cr_xMoO_{20+x/2-\delta}$ ($x = 0$, 0.05, 0.1, 0.15, 0.2, 0.25, 0.35, 0.5) measured in static air. The $\sigma$ increases with an increase of temperature. At any constant temperature from 400 to 900 °C, $\sigma$ values of Cr-doped samples of $0.05 \leq x \leq 0.5$ are higher than that of the mother material ($x = 0$) (Fig. 2.3), indicating the conductivity enhancement by the $Cr^{6+}$ doping. At any constant temperature from 400 to 900 °C, the $x = 0.2$ composition exhibits the highest $\sigma$ value. Therefore, we focus on this $x = 0.2$ composition for further studies. The $\sigma$ of $Ba_7Nb_{3.8}Cr_{0.2}MoO_{20.1-\delta}$ is $1.6 \times 10^{-3}$ S cm$^{-1}$ at 508 °C and $1.1 \times 10^{-2}$ S cm$^{-1}$ at 904 °C. The $\sigma$ of the $x = 0.2$ sample is 3.9 times higher than that of the $x = 0$ sample at 500 °C. The higher conductivity is attributable to the larger amount of excess oxygen 0.1 in $Ba_7Nb_{3.8}Cr_{0.2}MoO_{20.1}$ ($x = 0.2$) than that 0.0 in $Ba_7Nb_4MoO_{20}$.

Figure 2.2b shows the oxygen partial pressure $P(O_2)$ dependence of $\sigma$ for $Ba_7Nb_{3.8}Cr_{0.2}MoO_{20.1-\delta}$ at 304 and 604 °C. $\sigma$ is almost independent of $P(O_2)$ in the $P(O_2)$ regions from 1 to $2.2 \times 10^{-27}$ atm at 304 °C and 1 to $1.5 \times 10^{-26}$ atm at 604 °C, indicating the wide electrolyte domain, extremely high chemical and electrical stability of $Ba_7Nb_{3.8}Cr_{0.2}MoO_{20.1-\delta}$. The XRD patterns of $Ba_7Nb_{3.8}Cr_{0.2}MoO_{20.1}$ before and after the measurements of $\sigma$ at different $P(O_2)$ are similar with each other, indicating the high chemical stability of this material (Fig. 2.4). The electrical conductivity of $Ba_7Nb_{3.8}Cr_{0.2}MoO_{20.1-\delta}$ in wet air $\sigma_{wet}$ is nearly equal to that under dry air $\sigma_{dry}$ (Fig. 2.5), which suggests less proton conduction. The proton transport number under wet air flow was calculated by the equation $t_{H^+} = (\sigma_{wet} - \sigma_{dry})/\sigma_{wet}$. The $t_{H^+}$ for the composition $x = 0.2$ is ranged from 0.007 to 0.10, which is much lower than that for $x = 0$ ($0.33 \leq t_{H^+} \leq 0.74$) [22] and$Ba_7Ta_{3.7}Mo_{1.3}O_{20.15-\delta}$ ($0.2 \leq t_{H^+} \leq 0.5$) [26]. The $t_{H^+}$ for the $x = 0.2$ is as low as $Ba_7Nb_{3.9}Mo_{1.1}O_{20.05}$ ($0.03 \leq t_{H^+} \leq 0.11$) [23]. The lower $t_{H^+}$ of $Ba_7Nb_{3.8}Cr_{0.2}MoO_{20.1-\delta}$in wet air flow

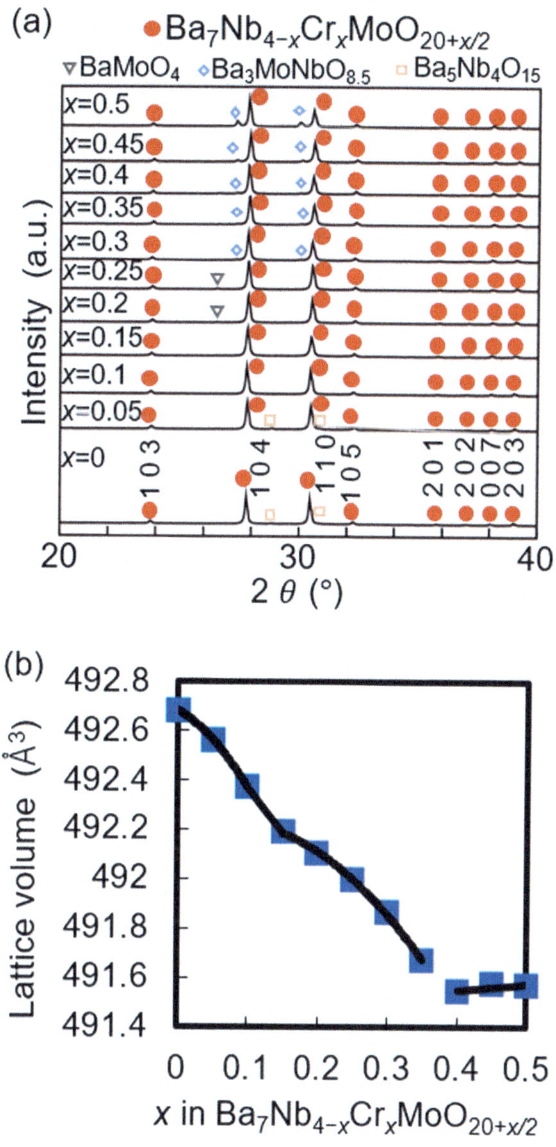


**Fig. 2.1** **a** XRD patterns of $Ba_7Nb_{4-x}Cr_xMoO_{20+x/2}$ ($x = 0, 0.05, 0.1, 0.15, 0.2, 0.25, 0.3, 0.35, 0.4, 0.45, 0.5$). Closed red circles, open black inverted triangles, open blue rhombuses, and open orange squares denote peak positions of trigonal $P\bar{3}m1$ 7H polytype $Ba_7Nb_{4-x}Cr_xMoO_{20+x/2}$, $BaMoO_4$, $Ba_3MoNbO_{8.5}$, and $Ba_5Nb_4O_{15}$, respectively. **b** Variation of the lattice volume of $Ba_7Nb_{4-x}Cr_xMoO_{20+x/2}$ ($x = 0, 0.05, 0.1, 0.15, 0.2, 0.25, 0.3, 0.35, 0.4, 0.45, 0.5$) with $x$ at RT. 

is attributable to the larger amount of excess oxygen atoms and smaller amount of oxygen vacancies. These results strongly suggest that the predominant conducting species in $Ba_7Nb_{3.8}Cr_{0.2}MoO_{20.1-\delta}$ is oxide ion.

**Table 2.1** Lattice parameters and lattice volumes of $Ba_7Nb_{4-x}Cr_xMoO_{20+x/2}$ ($x$ = 0, 0.05, 0.1, 0.15, 0.2, 0.25, 0.3, 0.35, 0.4, 0.45, 0.5) versus Cr content $x$ at RT

| Composition $x$ | Lattice parameters (Å) | | Lattice volume (Å$^3$) |
|---|---|---|---|
| | $a$ | $c$ | |
| 0.00 | 5.8659(3) | 16.5334(5) | 492.68(6) |
| 0.05 | 5.8641(3) | 16.5394(4) | 492.56(6) |
| 0.10 | 5.8640(3) | 16.5339(6) | 492.37(9) |
| 0.15 | 5.8622(5) | 16.5376(5) | 492.19(8) |
| 0.20 | 5.8610(5) | 16.5420(5) | 492.10(8) |
| 0.25 | 5.8598(4) | 16.5250(4) | 492.00(7) |
| 0.30 | 5.8594(7) | 16.5325(7) | 491.86(11) |
| 0.35 | 5.8572(5) | 16.5489(4) | 491.67(8) |
| 0.40 | 5.8573(6) | 16.5441(5) | 491.54(10) |
| 0.45 | 5.8587(8) | 16.5371(7) | 491.57(14) |
| 0.50 | 5.8574(12) | 16.5443(9) | 491.57(21) |

The number in the parenthesis is the estimated standard deviation of the last digit

To discuss the origin of high oxide-ion conductivity of $Ba_7Nb_{3.8}Cr_{0.2}MoO_{20.1-\delta}$, Rietveld refinements were performed using neutron-powder-diffraction data measured at 22 and 800 °C. The data were analyzed by the crystal structure of the $P\overline{3}m1$ 7H-polytype (Figs. 2.6, 2.7a; Tables 2.2, 2.3). In preliminary analyses, we examined the site preference of Cr cation in $Ba_7Nb_{3.8}Cr_{0.2}MoO_{20.1-\delta}$ at 22 and 800 °C. It was found that the reliability factor for the structural model where the Cr cation is placed at the $M2$ site ($R_{wp}$ = 6.21% at 22 °C and 6.27% at 800 °C) is lower than those at the $M1$ site ($R_{wp}$ = 6.46% at 22 °C and 6.37% at 800 °C) and $M3$ site ($R_{wp}$ = 6.35% at 22 °C and 6.31% at 800 °C). Therefore, the Cr dopant is preferentially located at the $M2$ site near the oxide-ion conducting c′ layer, which is attributable to the smaller volumes of $M2O_4$ tetrahedron (3.704 Å$^3$) and $M2O_6$ octahedron (10.645 Å$^3$) than those of the octahedra $M1O_6$ (10.858 Å$^3$) and $M3O_6$ (11.051 Å$^3$). The bond-valence sums (BVSs) calculated for the refined structure at RT are consistent with the average oxidation numbers (Table 2.2), which validates the refined crystal structure. It should be noted that the BVS of Cr at RT is estimated to be 5.9 valence unit, which strongly suggests the oxidation number of Cr is + 6 at RT. XPS spectra of the $x$ = 0.2 composition indicate that Ba, Mo and Nb adopt + 2, + 6 and + 5 oxidation states, respectively, at RT (Fig. 2.8). Therefore, the chemical formula is $(Ba^{2+})_7(Nb^{5+})_{3.8}(Cr^{6+})_{0.2}(Mo^{6+})(O^{2-})_{20.1}$ at RT and there is the excess oxygen atom $O_{0.1}$ in $Ba_7Nb_{3.8}Cr_{0.2}MoO_{20.1}$. The oxygen content 20.1 − $\delta$ in $Ba_7Nb_{3.8}Cr_{0.2}MoO_{20.1-\delta}$ at 800 °C was determined to be 20.005 using the TG analyses (Fig. 2.9), thus, the linear constraint $6g(O; O1) + 3g(O; O5) = 2.005$ was used in the Rietveld analysis of $Ba_7Nb_{3.8}Cr_{0.2}MoO_{20.1-\delta}$ at 800 °C (Table 2.3).

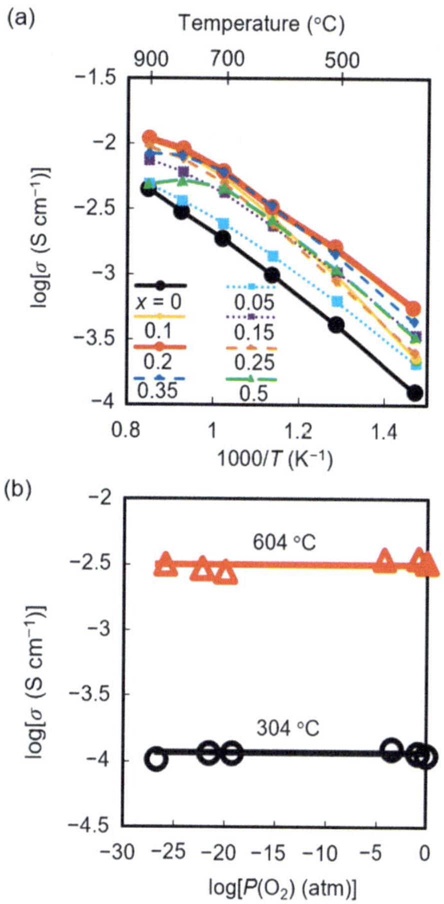


**Fig. 2.2** **a** Arrhenius plots of the electrical conductivities $\sigma$ of $Ba_7Nb_{4-x}Cr_xMoO_{20+x/2-\delta}$ ($x = 0, 0.05, 0.1, 0.15, 0.2, 0.25, 0.35, 0.5$) in static air from 300 to 900 °C. **b** Oxygen partial pressure $P(O_2)$ dependencies of the $\sigma$ for $Ba_7Nb_{3.8}Cr_{0.2}MoO_{20.1-\delta}$ at 304 and 604 °C. Copyright© 2022 Authors

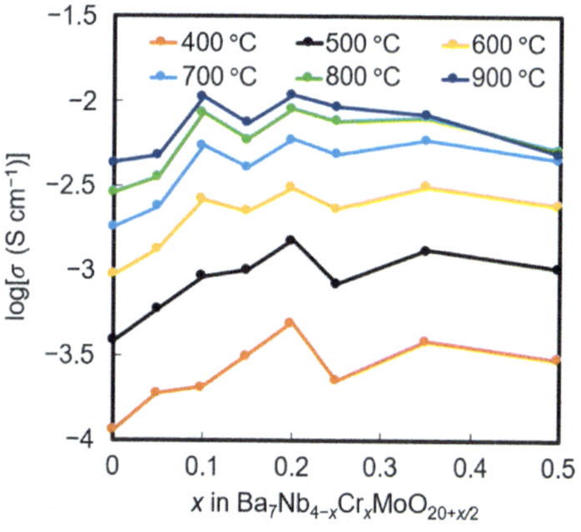


**Fig. 2.3** Cr content $x$ dependence of electrical conductivities $\sigma$ of $Ba_7Nb_{4-x}Cr_xMoO_{20+x/2-\delta}$ ($x = 0, 0.05, 0.1, 0.15, 0.2, 0.25, 0.35, 0.5$) in static air at constant temperatures from 300 to 900 °C. Electrical conductivity was calculated by interpolating the data in Fig. 2.2a. Copyright© 2022 Authors

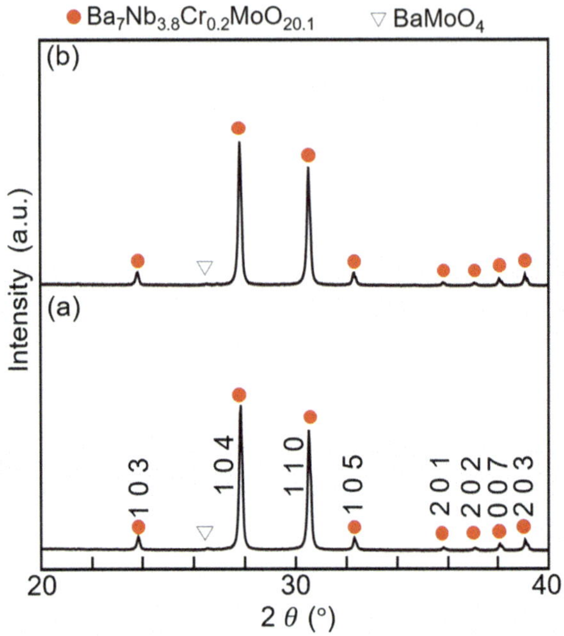


**Fig. 2.4** XRD patterns of $Ba_7Nb_{3.8}Cr_{0.2}MoO_{20.1}$ **a** before and **b** after the measurements of electrical conductivities at different oxygen partial pressure under the mixture of dry air, $O_2$, $N_2$, and/or 5% $H_2$ in $N_2$ flow (Fig. 2.2b). Copyright© 2022 Authors

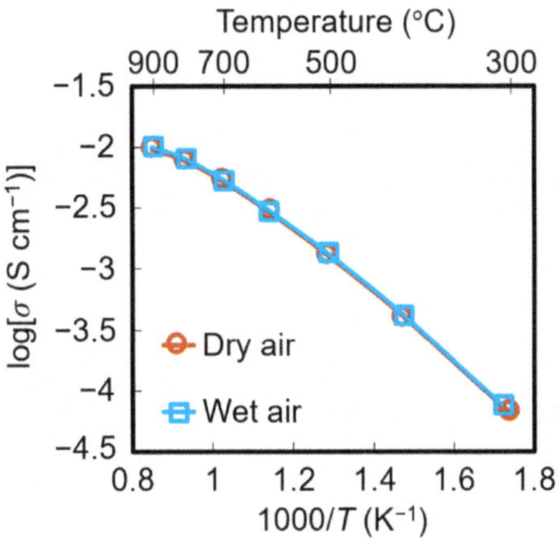


**Fig. 2.5** Arrhenius plots of the electrical conductivities of $Ba_7Nb_{3.8}Cr_{0.2}MoO_{20.1-\delta}$ in wet air (vapor partial pressure, $P(H_2O) =$ 0.021 atm; blue line and squares) and in dry air (red line and circles). Copyright© 2022 Authors

Figure 2.7b shows the refined crystal structure of $Ba_7Nb_{3.8}Cr_{0.2}MoO_{20.1-\delta}$ at 800 °C, which has hexagonal (h) and cubic (c) close-packed $BaO_3$ layers, and oxygen-deficient cubic (c′) close-packed $Ba(O1)_{2-y}(O5)_z$ layer where $y$ is the amount of oxygen vacancy at the O1 site and $z$ is the amount of interstitial oxygen atoms at an O5 site. The structure consists of $(Nb/Mo1)O_6$ and $(Nb/Mo3)O_6$ octahedra and $(Nb/Mo/Cr2)O_{4-y+z}$ polyhedra, and Ba cations.

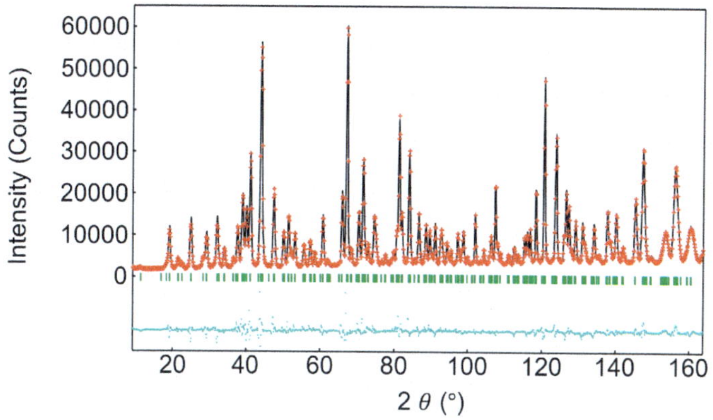


**Fig. 2.6** Rietveld pattern of $Ba_7Nb_{3.8}Cr_{0.2}MoO_{20.1}$ of neutron-diffraction data at 22 °C. The observed intensities, calculated intensities and difference pattern are shown by red marks, black line and light blue line, respectively. Green tick marks stand for calculated Bragg peak positions. Copyright© 2022 Authors

Oxide-ion diffusion pathways of $Ba_7Nb_{3.8}Cr_{0.2}MoO_{20.1-\delta}$ were experimentally examined by the neutron scattering length density distribution (NSLD) obtained by MEM analysis of neutron-diffraction data taken at 800 °C. $Ba_7Nb_{3.8}Cr_{0.2}MoO_{20.1-\delta}$ shows a connected NSLDs between the lattice tetrahedral O1 and interstitial octahedral O5 sites (Fig. 2.10a), which indicates the oxide-ion O1–O5 diffusion pathway. The MEM NSLD shows that oxide ion migrates two-dimensionally, which is supported by bond-valence-based energy landscape (Fig. 2.10b). Similar oxide-ion diffusion pathways were observed in MEM NSLD distribution of $Ba_7Nb_{3.9}Mo_{1.1}O_{20.05}$, [23] $Ba_3WNbO_{8.5-\delta}$, [29] and $Ba_3MoNbO_{8.5-\delta}$ [42] and suggested in BVE landscapes of $Ba_7Nb_4MoO_{20}$, [24] $Ba_7Ta_{3.7}Mo_{1.3}O_{20.15}$, [26] $Ba_3WNbO_{8.5-\delta}$, [29] and $Ba_3MoNbO_{8.5-\delta}$ [42]. Therefore, the oxide-ion tetrahedral oxygen–octahedral oxygen diffusion in the c′ layer is a common feature of oxide-ion conductors based on $Ba_7Nb_4MoO_{20}$ and $Ba_3MoNbO_{8.5}$. The presence of the interstitial O5 atoms and the O1–O5 diffusion pathway is the structural origin of the high oxide-ion conductivity. The O1–O5 interatomic distance 1.426(3) Å is short and the oxide ions migrate through both interstitial O5 and lattice O1 sites, therefore, the oxide ions migrate via an interstitialcy diffusion mechanism.

The small BVS value at *M*1 site 4.53 suggests that occupancy factor of Nb at the *M*1 site nearly equals to unity, suggesting that the oxidation number of *M*1 atom is nearly equals to 5, which agree with the calculated average BVS value of 4.53 within 10%. Thus, the average BVS value is meaningful.

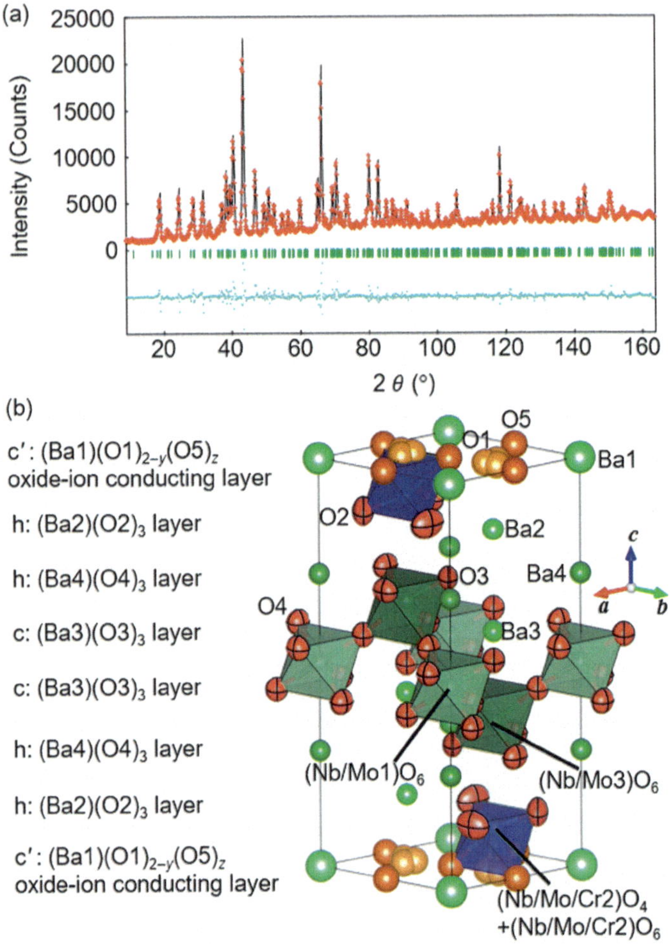


**Fig. 2.7** Structure analysis of $Ba_7Nb_{3.8}Cr_{0.2}MoO_{20.1-\delta}$ at 800 °C. **a** Rietveld pattern of neutron-diffraction data measured in situ at 800 °C. Red crosses, black lines, green bars, and light blue dots represent observed intensities, calculated intensities, peak positions, and difference plots, respectively. **b** Refined crystal structure of $Ba_7Nb_{3.8}Cr_{0.2}MoO_{20.1-\delta}$. Thermal ellipsoids are drawn at the 90% probability level. Here, $(Nb/Mo/Cr_2)O_4$ + $(Nb/Mo/Cr_2)O_6$ denotes the hybrid of $3g$(O; O1)·$(Nb/Mo/Cr_2)O_4$ tetrahedron and $g$(O; O5)·$(Nb/Mo/Cr_2)O_6$ octahedron (= $(Nb/Mo/Cr_2)O_{4\text{-}y+z}$). Copyright© 2022 Authors

**Table 2.2** Refined crystallographic parameters and reliability factors in the Rietveld analysis of the neutron-diffraction data of $Ba_7Nb_{3.8}Cr_{0.2}MoO_{20.1}$ at 22 °C

| Site *s* atom label | Atom *X* | Wyckoff position | *g*(*X*; *s*) | *x* | *y* | *z* | $U_{iso}$ (Å$^2$) | BVS (vu) |
|---|---|---|---|---|---|---|---|---|
| Ba1 | Ba | 1*a* | 1 | 0 | 0 | 0 | 0.016(2) | 2.03 |
| Ba2 | Ba | 2*d* | 1 | 1/3 | 2/3 | 0.8249(4) | 0.0072(7) | 2.26 |
| Ba3 | Ba | 2*d* | 1 | 1 | 2/3 | 0.5742(4) | 0.0072(7) | 2.38 |
| Ba4 | Ba | 2*c* | 1 | 0 | 0 | 0.2806(4) | 0.0072(7) | 1.92 |
| Nb/Mo1 | $Nb_{0.7917}Mo_{0.2083}$ | 1*b* | 1 | 0 | 0 | 1/2 | 0.0059 (5) | 4.53 |
| Nb/Mo/Cr2 | $Nb_{0.7125}Mo_{0.1875}Cr_{0.1000}$ | 2*d* | 1 | 1/3 | 2/3 | 0.0956(3) | 0.0059(5) | 5.77 |
| Nb/Mo3 | $Nb_{0.7917}Mo_{0.2083}$ | 2*d* | 1 | 1/3 | 2/3 | 0.3498(3) | 0.0059(5) | 4.69 |
| O1 | O | 6*d* | 0.332(5) | 0.3555(17) | 0.7109(3) | − 0.0118(5) | 0.022(4) | 1.86 |
| O2 | O | 6*i* | 1 | 0.1678(6) | − 0.1678(6) | 0.13086(18) | 0.0144(8) | 2.10 |
| O3 | O | 6*i* | 1 | 0.1643(6) | − 0.1643(6) | 0.4319(2) | 0.087(7) | 1.91 |
| O4 | O | 6*i* | 1 | 0.4951(4) | − 0.4951(4) | 0.29482(17) | 0.082(7) | 2.07 |
| O5 | O | 3*e* | 0.035(10) | 1/2 | 0 | 0 | 0.022(4) | 1.08 |

Crystal system: trigonal. Space group: $P\bar{3}m1$. Lattice parameters: $a = b = 5.85783(17)$ Å, $c = 16.5388(4)$ Å. Number of formula per unit cell: $Z = 1$
$g(X; s)$: Occupancy factor of $X$ atom at the $s$ site. Linear constraints: $g(\text{Ba; Ba1}) = g(\text{Ba; Ba2}) = g(\text{Ba; Ba3}) = g(\text{Ba; Ba4}) = g(Nb_{0.7917}Mo_{0.2083}; \text{Mo/Nb1}) = g(Nb_{0.7125}Mo_{0.1875}Cr_{0.1000}; \text{Mo/Nb/Cr2}) = g(Nb_{0.7917}Mo_{0.2083}; \text{Mo/Nb3}) = g(\text{O; O2}) = g(\text{O; O3}) = g(\text{O; O4}) = 1; 6g(\text{O; O1}) + 3g(\text{O; O5}) = 2.1$
$U_{iso}(Xn)$: Isotropic atomic displacement parameter of $X$ atom at the $Xn$ site. Linear constraints: $U_{iso}(\text{Ba2}) = U_{iso}(\text{Ba3}) = U_{iso}(\text{Ba4})$, $U_{iso}(\text{Mo/Nb1}) = U_{iso}(\text{Mo/Nb2}) = U_{iso}(\text{Mo/Nb3})$, $U_{iso}(\text{O1}) = U_{iso}(\text{O5})$. Reliability factors: $R_{wp} = 6.21\%$, $R_p = 4.85\%$, $R_B = 1.76\%$, $R_F = 0.76\%$
BVS: bond valence sum. The BVS value for each $Mi$ site ($i = 1$, 2 and 3). BVS($Mi$) is the average value of BVSs of Nb, Mo and Cr cations, which was calculated using the following equation. $\text{BVS}(Mi) = \text{BVS}(\text{Nb}; Mi) \times g(\text{Nb}; Mi) + \text{BVS}(\text{Mo}; Mi) \times g(\text{Mo}; Mi) + \text{BVS}(\text{Cr}; Mi) \times g(\text{Cr}; Mi)$. ($i = 1, 2, 3$) Eq. (2.1). Here, BVS($X$; $Mi$) and $g(X; Mi)$ denote the BVS of $X$ atom at the $Mi$ site and occupancy factor of $X$ atom at the $Mi$ site, respectively. The BVS($X$; $Mi$) was calculated as follows. $\text{BVS}(X; Mi) = \Sigma_k g(\text{O}_k)\{\exp([R(X; \text{O}) - d(X - \text{O}_k)]/b(X; \text{O}))\}$ Eq. (2.2). where the $\text{O}_k$ is the $k$th oxygen site in the $Mi(\text{O}_k)_n$ polyhedron ($n = 6$ for $i = 1$ and 3, $n = 9$ for $i = 2$), $g(\text{O}_k)$ is the occupancy factor of oxygen atom for the $k$th oxygen site, $d(X - \text{O}_k)$ is the distance between the central cation $X$ and $k$th oxygen site, $R(X; \text{O})$ is the bond-valence parameter of cation $X$ for oxygen atom, and $b(X; \text{O})$ is the empirical parameter of cation $X$ for oxygen atom. We used the $R(X; \text{O})$ and $b(X; \text{O})$ values after Gagné and Hawthorne [41]. Since the peaks of impurity $BaMoO_4$ were not detected in the neutron diffraction pattern, Rietveld analysis was performed by the single phase

**Table 2.3** Refined crystallographic parameters and reliability factors in the Rietveld analysis of the neutron-diffraction data of $Ba_7Nb_{3.8}Cr_{0.2}MoO_{20.1-\delta}$ at 800 °C

| Site *s* Atom label | Atom *X* | Wyckoff position | *g*(*X*; *s*) | *x* | *y* | *z* |
|---|---|---|---|---|---|---|
| Ba1 | Ba | 1*a* | 1 | 0 | 0 | 0 |
| Ba2 | Ba | 2*d* | 1 | 1/3 | 2/3 | 0.8248(7) |
| Ba3 | Ba | 2*d* | 1 | 1 | 2/3 | 0.5745(9) |
| Ba4 | Ba | 2*c* | 1 | 0 | 0 | 0.2832(8) |
| Nb/Mo1 | $Nb_{0.7917}Mo_{0.2083}$ | 1*b* | 1 | 0 | 0 | 1/2 |
| Nb/Mo/Cr2 | $Nb_{0.7125}Mo_{0.1875}Cr_{0.1000}$ | 2*d* | 1 | 1/3 | 2/3 | 0.0939(6) |
| Nb/Mo3 | $Nb_{0.7917}Mo_{0.2083}$ | 2*d* | 1 | 1/3 | 2/3 | 0.3499(6) |
| O1 | O | 6*d* | 0.317(7) | 1/3 | 2/3 | −0.0121(9) |
| O2 | O | 6*i* | 1 | 0.1689(14) | −0.1689(14) | 0.1300(5) |
| O3 | O | 6*i* | 1 | 0.1659(12) | −0.1659(12) | 0.4335(5) |
| O4 | O | 6*i* | 1 | 0.4925(10) | −0.4925(10) | 0.2957(4) |
| O5 | O | 3*e* | 0.034(15) | ½ | 0 | 0 |

| Site *s* Atom label | $U_{iso}$ or $U_{eq}$ (Å$^2$) | $U_{11}$ (Å$^2$) | $U_{22}$ (Å$^2$) | $U_{33}$ (Å$^2$) | $U_{12}$ (Å$^2$) | $U_{13}$ (Å$^2$) | $U_{23}$ (Å$^2$) | BVS (vu) |
|---|---|---|---|---|---|---|---|---|
| Ba1 | 0.062(9) | | | | | | | 1.90 |
| Ba2 | 0.0279(15) | | | | | | | 2.01 |
| Ba3 | 0.0279(15) | | | | | | | 2.24 |
| Ba4 | 0.0279(15) | | | | | | | 1.73 |
| Nb/Mo1 | 0.0210(9) | | | | | | | 4.28 |
| Nb/Mo/Cr2 | 0.0210(9) | | | | | | | 5.49 |
| Nb/Mo3 | 0.0210(9) | | | | | | | 4.67 |

(continued)

**Table 2.3** (continued)

| Site $s$ Atom label | $U_{iso}$ or $U_{eq}$ (Å$^2$) | $U_{11}$ (Å$^2$) | $U_{22}$ (Å$^2$) | $U_{33}$ (Å$^2$) | $U_{12}$ (Å$^2$) | $U_{13}$ (Å$^2$) | $U_{23}$ (Å$^2$) | BVS (vu) |
|---|---|---|---|---|---|---|---|---|
| O1 | 0.044(8) | | | | | | | 1.79 |
| O2 | 0.051(5) | 0.064(5) | 0.064(5) | 0.049(5) | 0.051(7) | 0 | 0 | 1.98 |
| O3 | 0.033(3) | 0.033(3) | 0.033(3) | 0.037(4) | 0.019(4) | 0 | 0 | 1.79 |
| O4 | 0.033(4) | 0.038(4) | 0.038(4) | 0.035(4) | 0.028(4) | 0 | 0 | 1.98 |
| O5 | 0.044(8) | | | | | | | 1.07 |

Crystal system: trigonal. Space group: $P\overline{3}m1$. Lattice parameters: $a = b = 5.92266(17)$ Å, $c = 16.7055(6)$ Å. Number of formula per unit cell: $Z = 1$

$g(X; s)$: Occupancy factor of $X$ atom at the $s$ site. Linear constraints: $g$(Ba; Ba1) = $g$(Ba; Ba2) = $g$(Ba; Ba3) = $g$(Ba; Ba4) = $g$($Nb_{0.7917}Mo_{0.2083}$; Mo/Nb1) = $g$($Nb_{0.7125}Mo_{0.1875}Cr_{0.1000}$; Mo/Nb/Cr2) = $g$($Nb_{0.7917}Mo_{0.2083}$; Mo/Nb3) = $g$(O; O2) = $g$(O; O3) = $g$(O; O4) = 1; 6$g$(O; O1) + 3$g$(O; O5) = 2.005

$U_{iso}(Xn)$: Isotropic atomic displacement parameter of $X$ atom at the $Xn$ site. Linear constraints: $U_{iso}$(Ba2) = $U_{iso}$(Ba3) = $U_{iso}$(Ba4), $U_{iso}$(Mo/Nb1) = $U_{iso}$(Mo/Nb2) = $U_{iso}$(Mo/Nb3), $U_{iso}$(O1) = $U_{iso}$(O5). Reliability factors: $R_{wp} = 6.27\%$, $R_p = 4.65\%$, $R_B = 2.26\%$, $R_F = 1.06\%$

BVS: bond valence sum. BVSs were calculated using the bond-valence parameters reported by Gagné and Hawthorne. Since the peaks of impurity $BaMoO_4$ were not detected in the neutron diffraction pattern, Rietveld analysis was performed by the single phase. The BVS value for each $Mi$ site ($i$ = 1, 2 and 3). BVS($Mi$) is the average value of BVSs of Nb, Mo and Cr cations, which was calculated using the following equation. BVS($Mi$) = BVS(Nb; $Mi$) × $g$(Nb; $Mi$) + BVS(Mo; $Mi$) × $g$(Mo; $Mi$) + BVS(Cr; $Mi$) × $g$(Cr; $Mi$). ($i$ =1, 2, 3) Eq. (2.1). Here, BVS($X$; $Mi$) and $g$($X$; $Mi$) denote the BVS of $X$ atom at the $Mi$ site and occupancy factor of X atom at the $Mi$ site, respectively. The BVS($X$; $Mi$) was calculated as follows. BVS($X$; $Mi$) = $\Sigma_k g(O_k)\{\exp([R(X; O)-d(X-O_k)]/b(X; O))\}$ Eq. (2.2). where the $O_k$ is the $k$th oxygen site in the $Mi(O_k)_n$ polyhedron ($n$ = 6 for $i$ = 1 and 3, $n$ = 9 for $i$ = 2), $g(O_k)$ is the occupancy factor of oxygen atom for the $k$th oxygen site, $d(X - O_k)$ is the distance between the central cation $X$ and $k$th oxygen site, $R(X; O)$ is the bond-valence parameter of cation $X$ for oxygen atom, and $b(X; O)$ is the empirical parameter of cation $X$ for oxygen atom. We used the $R(X; O)$ and $b(X; O)$ values after Gagné and Hawthorne [41].

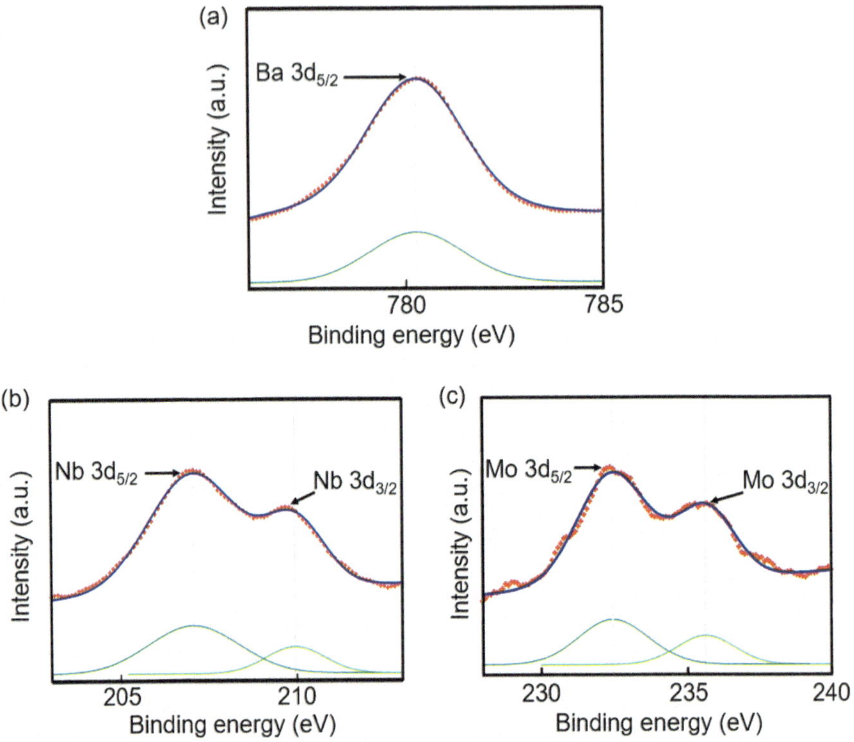


**Fig. 2.8** XPS spectra and fitting curves of **a** Ba 3d, **b** Nb 3d, and **c** Mo 3d states of $Ba_7Nb_{3.8}Cr_{0.2}MoO_{20.1}$. Red crosses and blue lines denote the observed and calculated intensities, respectively. Dark green and light green curves stand for the calculated intensities for $3d_{5/2}$ and $3d_{3/2}$ components, respectively. By the fitting, we obtained the following binding energies, which agree well with the reported values described in the parentheses. Ba $3d_{5/2}$: 780.3 eV (780 eV) [38], $Nb3d_{5/2}$: 207.1 eV (207.2 eV) [39], $Nb3d_{3/2}$: 210.0 eV (210 eV) [39], $Mo3d_{5/2}$: 232.5 eV (232.3–232.6 eV) [40], $Mo3d_{3/2}$: 235.6 eV (235.4–235.7 eV) [40]. The obtained and reported values are close, thus, the valences of Nb and Mo are 5+ and 6+, respectively. Copyright© 2022 Authors

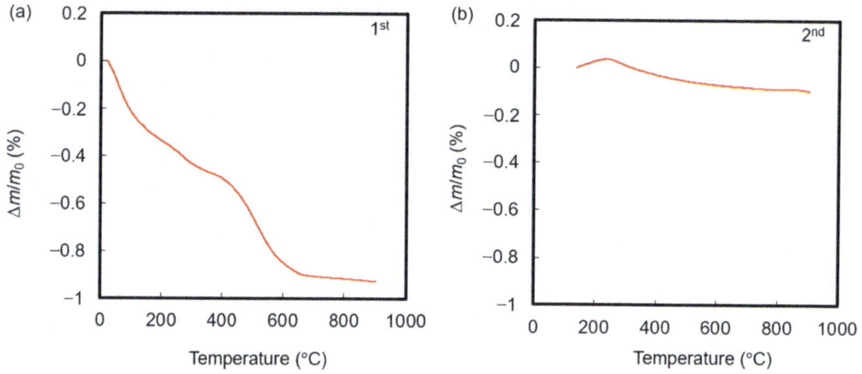


**Fig. 2.9** Thermogravimetric (TG) curves of $Ba_7Nb_{3.8}Cr_{0.2}MoO_{20.1-\delta}$ on heating in dry air in **a** 1st and **b** 2nd cycles. Here, $\Delta m/m_0$ (%) = 100 $(m - m_0)/m_0$. The $m$ is the sample weight at a temperature, and $m_0$ represents the sample weight at 20 °C (a) and 141 °C (b). Copyright© 2022 Authors

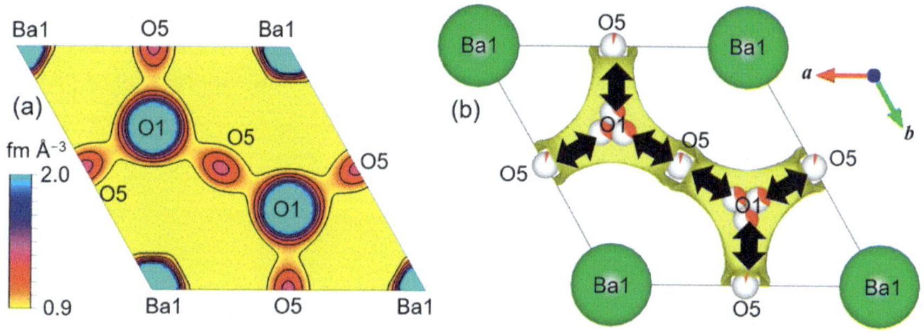


**Fig. 2.10** **a** MEM neutron scattering length density distribution on the *ab* plane at $z = 0$ of $Ba_7Nb_{3.8}Cr_{0.2}MoO_{20.1-\delta}$ at 800 °C. The contour lines are from 0.8 to 2 fm $Å^{-3}$ by the step of 0.2 fm $Å^{-3}$. **b** Bond valence-based energy landscapes and refined crystal structure in the oxygen-deficient c′ layer of $Ba_7Nb_{3.8}Cr_{0.2}MoO_{20.1-\delta}$ at 800 °C. Yellow isosurface level: 2.1 eV. Copyright© 2022 Authors

## 2.4 Conclusion

In conclusion, we have demonstrated simultaneous improvement of oxide-ion conductivity and suppression of proton conduction by $Cr^{6+}$ doping into the hexagonal perovskite related oxide $Ba_7Nb_4MoO_{20}$. We have synthesized new hexagonal perovskite-related oxides $Ba_7Nb_{4-x}Cr_xMoO_{20+x/2}$ ($x = 0.05, 0.1, 0.15, 0.2, 0.25, 0.3, 0.35, 0.4, 0.45, 0.5$). The lattice volume linearly decreases with an increase of Cr content $x$ in $Ba_7Nb_{4-x}Cr_xMoO_{20+x/2}$, indicating the formation of solid solutions up to the solubility limit at $x = 0.394(14)$. Among the $Ba_7Nb_{4-x}Cr_xMoO_{20+x/2-\delta}$ ($x = 0.0, 0.05, 0.1, 0.15, 0.2, 0.25, 0.35, 0.5$), $Ba_7Nb_{3.8}Cr_{0.2}MoO_{20.1-\delta}$ shows the highest electrical conductivity of $1.6 \times 10^{-3}$ S cm$^{-1}$ at 508 °C and $1.1 \times 10^{-2}$ S cm$^{-1}$ at 904 °C in static air. The electrical conductivity of $Ba_7Nb_{3.8}Cr_{0.2}MoO_{20.1}$ is almost constant regardless of the oxygen partial pressure $P(O_2)$ in the $P(O_2)$ regions from 1 to $2.2 \times 10^{-27}$ atm at 304 °C and 1 to $1.5 \times 10^{-26}$ atm at 604 °C, indicating the wide electrolyte domains, extremely high chemical and electrical stability, and predominant oxide-ion conduction in dry conditions. The proton transport number of $Ba_7Nb_{3.8}Cr_{0.2}MoO_{20.1-\delta}$ under wet conditions is ranged from 0.007 to 0.10, which is much smaller than that of the mother material $Ba_7Nb_4MoO_{20-\delta}$ (0.33–0.74), which indicates that the $Cr^{6+}$ doping considerably suppresses the proton conduction compared with the mother material. The structure analyses have shown that $Ba_7Nb_{3.8}Cr_{0.2}MoO_{20.1-\delta}$ is a hexagonal perovskite related oxide at 22 and 800 °C. Refined crystal structure of $Ba_7Nb_{3.8}Cr_{0.2}MoO_{20.1}$ has oxygen-deficient cubic (c′) close-packed $Ba(O1)_{2-y}(O5)_z$ layer where $y$ is the amount of oxygen vacancy at the tetrahedral O1 site and $z$ is the amount of interstitial octahedral oxygen atoms at an O5 site. The $Cr^{6+}$ cation is located at the $M2$ site near the c′ layer, which leads to the excess oxygen and high conductivity and is likely to suppress the proton conduction. MEM analyses have demonstrated that oxide ions two-dimensionally migrate through the lattice O1 and interstitial O5 sites in the c′ layer via the interstitialcy diffusion mechanism at 800 °C, which enables the high oxide-ion conduction. The $Cr^{6+}$ doping in various hexagonal perovskite-related oxides would be a new strategy for the simultaneous improvement of oxide-ion conductivity and suppression of proton conduction.

## References


1. Ishihara T, Matsuda H, Takita Y (1994) Doped $LaGaO_3$ perovskite type oxide as a new oxide ionic conductor. J Am Chem Soc 116(9):3801–3803
2. Ishihara T, Kilner JA, Honda M, Takita Y (1997) Oxygen surface exchange and diffusion in the new perovskite oxide ion conductor $LaGaO_3$. J Am Chem Soc 119(11):2747–2748
3. Huang K, Tichy RS, Goodenough JB (1998) Superior perovskite oxide-ion conductor; strontium- and magnesium-doped $LaGaO_3$: I, Phase relationships and electrical properties. J Am Ceram Soc 81(10):2565–2575
4. Takai S, Sugiura K, Esaka T (1999) Ionic conduction properties of $Pb_{1-x}M_xWO_{4+\delta}$ (M = Pr, Tb). Mater Res Bull 34(2):193–202
5. Takai S, Touda S, Oikawa K, Mori K, Torii S, Kamiyama T, Esaka T (2002) Powder neutron diffraction study of Ln-substituted $PbWO_4$ oxide ion conductors. Solid State Ion 148(1–2):123–133
6. Vannier RN, Skinner SJ, Chater RJ, Kilner JA, Mairesse G (2003) Oxygen transfer in BIMEVOX materials. Solid State Ion 160(1–2):85–92
7. Kreuer K (2003) Proton-conducting oxides. Annu Rev Mater Res 33:333–359
8. León-Reina L, Losilla ER, Martínez-Lara M, Martín-Sedeño MC, Bruque S, Núñez P, Sheptyakov DV, Aranda MAG (2005) High oxide ion conductivity in Al-doped germanium oxyapatite. Chem Mater 17(3):596–600
9. He T, Huang Y, He Q, Ji Y, Pei L, Liu J, Lu Z (2005) The effects on the structures and properties in the oxide-ion conductor $La_2Mo_2O9$ by partial substituting Ba for La. J Alloys Compd 388(1):145–152
10. Malavasi L, Kim HJ, Billinge SJL, Proffen T, Tealdi C, Flor G (2007) Nature of the monoclinic to cubic phase transition in the fast oxygen ion conductor $La_2Mo_2O_9$ (LAMOX). J Am Chem Soc 129(21):6903–6907

11. Kuang X, Green MA, Niu H, Zajdel P, Dickinson C, Claridge JB, Jantsky L, Rosseinsky MJ (2008) Interstitial oxide ion conductivity in the layered tetrahedral network melilite structure. Nat Mater 7(6):498–504
12. Quarez E, Noirault S, Caldes MT, Joubert O (2010) Water incorporation and proton conductivity in titanium substituted barium indate. J Power Sources 195(4):1136–1141
13. López CA, Pedregosa JC, Lamas DG, Alonso JA (2014) The strongly defective double perovskite $Sr_{11}Mo_4O_{23}$: crystal structure in relation to ionic conductivity. J Appl Crystallogr 47(4):1395–1401
14. Li M, Pietrowski MJ, De Souza RA, Zhang H, Reaney IM, Cook SN, Kilner JA, Sinclair DC (2014) A family of oxide ion conductors based on the ferroelectric perovskite $Na_{0.5}Bi_{0.5}TiO_3$. Nat Mater 13(1):31–35
15. Fop S, Skakle JMS, McLaughlin AC, Connor PA, Irvine JTS, Smith RI, Wildman EJ (2016) Oxide ion conductivity in the hexagonal perovskite derivative $Ba_3MoNbO_{8.5}$. J Am Chem Soc 138(51):16764–16769
16. Fujii K, Yashima M, Hibino K, Shiraiwa M, Fukuda K, Nakayama S, Ishizawa N, Hanashima T, Ohhara T (2018) High oxide-ion conductivity in Si-deficient $La_{9.565}(Si_{5.826}\square_{0.174})O_{26}$ apatite without interstitial oxygens due to the overbonded channel oxygens. J Mater Chem A 6(23):10835–10846
17. McCombie KS, Wildman EJ, Fop S, Smith RI, Skakle JMS, McLaughlin AC (2018) The crystal structure and electrical properties of the oxide ion conductor $Ba_3WNbO_{8.5}$. J Mater Chem A Mater 6(13):5290–5295
18. Yasui Y, Niwa E, Matsui M, Fujii K, Yashima M (2019) Discovery of a rare-earth-free oxide-ion conductor $Ca_3Ga_4O_9$ by screening through bond valence-based energy calculations, synthesis, and characterization of structural and transport properties. Inorg Chem 58(14):9460–9468
19. Auckett JE, Milton KL, Evans IR (2019) Cation distributions and anion disorder in $Ba_3NbMO_{8.5}$ (M = Mo, W) materials: implications for oxide ion conductivity. Chem Mater 31(5):1715–1719
20. Murakami T, Hester JR, Yashima M (2020) High proton conductivity in $Ba_5Er_2Al_2ZrO_{13}$, a hexagonal perovskite-related oxide with intrinsically oxygen-deficient layers. J Am Chem Soc 142(27):11653–11657
21. Zhang W, Fujii K, Niwa E, Hagihala M, Kamiyama T, Yashima M (2020) Oxide-ion conduction in the Dion-Jacobson phase $CsBi_2Ti_2NbO_{10-\delta}$. Nat Commun 11(1):1224
22. Fop S, McCombie KS, Wildman EJ, Skakle JMS, Irvine JTS, Connor PA, Savaniu C, Ritter C, Mclaughlin AC (2020) High oxide ion and proton conductivity in a disordered hexagonal perovskite. Nat Mater 19(7):752–757
23. Yashima M, Tsujiguchi T, Sakuda Y, Yasui Y, Zhou Y, Fujii K, Torii S, Kamiyama T, Skinner SJ (2021) High oxide-ion conductivity through the interstitial oxygen site in $Ba_7Nb_4MoO_{20}$-based hexagonal perovskite related oxides. Nat Commun 12:556
24. Fop S, Dawson JA, Fortes AD, Ritter C, McLaughlin AC (2021) Hydration and ionic conduction mechanisms of hexagonal perovskite derivatives. Chem Mater 33(12):4651–4660
25. Fop S (2021) Solid oxide proton conductors beyond perovskites. J Mater Chem A 9(35):18836–18856
26. Murakami T, Shibata T, Yasui Y, Fujii K, Hester JR, Yashima M (2021) High oxide-ion conductivity in a hexagonal perovskite-related oxide $Ba_7Ta_{3.7}Mo_{1.3}O_{20.15}$ with cation site preference and interstitial oxide ions. Small 18(10):2106785
27. Yaguchi H, Fujii K, Tsuchiya Y, Ogino H, Tsujimoto Y, Yashima M (2022) Ruddlesden-popper oxychlorides $Ba_3Y_2O_5Cl_2$, $Sr_3Sc_2O_5Cl_2$, and $Sr_2ScO_3Cl$: first examples of oxide-ion-conducting oxychlorides. ACS Appl Energy Mater 5(1):295–304
28. Saito K, Fujii K, Yashima M (2022) Oxide-ion and proton conductivity of the ordered perovskite $BaY_{1/3}Ga_{2/3}O_{2.5}$. J Solid State Chem 306:122733
29. Yasui Y, Tsujiguchi T, Sakuda Y, Hester JR, Yashima M (2022) Oxide-ion occupational disorder, diffusion path, and conductivity in hexagonal perovskite derivatives $Ba_3WNbO_{8.5}$ and $Ba_3MoNbO_{8.5}$. J Phys Chem C 126(5):2383–2393

30. Avdeev M, Hester JR (2018) ECHIDNA: a decade of high-resolution neutron powder diffraction at OPAL. J Appl Crystallogr 51(6):1597–1604
31. Izumi F, Momma K (2007) Three-dimensional visualization in powder diffraction. Solid State Phenom 130:15–20
32. Momma K, Ikeda T, Belik AA, Izumi F (2013) Dysnomia, a computer program for maximum-entropy method (MEM) analysis and its performance in the MEM-based pattern fitting. Powder Diffr 28(3):184–193
33. Adams S (2000) Modelling ion conduction pathways by bond valence pseudopotential maps. Solid State Ion 136–137:1351–1361
34. Chen H, Wong LL, Adams S (2019) SoftBV—a software tool for screening the materials genome of inorganic fast ion conductors. Acta Crystallogr Sect B Struct Sci Cryst Eng Mater 75(1):18–33
35. Momma K, Izumi F (2011) VESTA 3 for three-dimensional visualization of crystal, volumetric and morphology data. J Appl Crystallogr 44(6):1272–1276
36. García-González E, Parras M, González-Calbet JM (1999) Crystal structure of an unusual polytype: 7H-$Ba_7Nb_4MoO_{20}$. Chem Mater 11(2):433–437
37. Shannon RD (1976) Revised effective ionic radii and systematic studies of interatomic distances in halides and chalcogenides. Acta Crystallogr Sect A 32(5):751–767
38. Wegmann M, Watson L, Hendry A (2004) XPS analysis of submicrometer barium titanate powder. J Am Ceram Soc 87(3):371–377
39. Choi JG, Thompson LT (1996) XPS study of As-prepared and reduced molybdenum oxides. Appl Surf Sci 93(2):143–149
40. Yu J, Li Y, Xu X, Duan G, Li Y, Zhou W (2021) Rambutan-like $Nb_2O_5$@SHCs microspheres for improved microwave absorption performance. Compos Commun 24:100643
41. Gagné OC, Hawthorne FC (2015) Comprehensive derivation of bond-valence parameters for ion pairs involving oxygen. Acta Crystallogr Sect B Struct Sci Cryst Eng Mater 71(5):562–578
42. Yashima M, Tsujiguchi T, Fujii K, Niwa E, Nishioka S, Hester JR, Maeda K (2018) Direct evidence for two-dimensional oxide-ion diffusion in the hexagonal perovskite-related oxide $Ba_3MoNbO_{8.5-\delta}$. J Mater Chem A 7(23):13910–13916

# Chapter 3
# Dimer-Mediated Cooperative Mechanism of Ultrafast-Ion Conduction in Hexagonal Perovskite-Related Oxides

**Abstract** Oxide-ion and proton conductors have found diverse applications such as electrolytes of solid-oxide, proton-conducting, and hybrid-ion fuel cells. Research of fuel cells with higher energy efficiency at lower operating temperature has stimulated the search for ion conductors and improved and the study of ion diffusion mechanism. Ion conduction in hexagonal perovskite-related materials is rare, and the mechanism of ion diffusion is unclear. Herein, we report high oxide-ion and proton conductivity (bulk conductivities in wet air: 11 and 2.7 mS cm$^{-1}$ at 537 and 326 °C, respectively), high chemical and electrical stability in a new hexagonal perovskite-related oxide $Ba_7Nb_{3.8}Mo_{1.2}O_{20.1}$. Total direct current conductivity at 400 °C in wet air of $Ba_7Nb_{3.8}Mo_{1.2}O_{20.1}$ was 13 times higher than that of $Ba_7Nb_4MoO_{20}$. We also report a unique dimer-mediated cooperative mechanism of the high oxide-ion conduction of $Ba_7Nb_{3.8}Mo_{1.2}O_{20.1}$ (bulk conductivities in dry air: 10 mS cm$^{-1}$ at 593 °C and 1.1 mS cm$^{-1}$ at 306 °C). Ab initio molecular dynamics (AIMD) simulation, neutron-diffraction experiments at 800 °C, and neutron scattering length density analyses of $Ba_7Nb_{3.8}Mo_{1.2}O_{20.1}$ indicated that the excess oxygen atoms are incorporated by the formation of both fivefold coordinated $(Nb/Mo)O_5$ monomer and its $(Nb/Mo)_2O_9$ dimer with a corner-sharing oxygen atom and that the breaking and reforming of the dimers lead to the high oxide-ion conduction in the oxygen-deficient $BaO_{2.1}$ c′ layer. The long distance between Nb/Mo and Ba cations sandwiching the c′ layer of $Ba_7Nb_{3.8}Mo_{1.2}O_{20.1}$ was found to be responsible for its low activation energy for oxide-ion conduction, leading to high conductivity at low temperatures. AIMD simulations showed that high proton conduction is attributable to proton migration in the hexagonal close-packed $BaO_3$ layers of $Ba_7Nb_{3.8}Mo_{1.2}O_{20.1}$, which is supported by its crystal structure analysis. The present findings hold a great promise for the development and design of ion conductors.



Y. Sakuda, *Discovery of* $Ba_7Nb_4MoO_{20}$*-Based Materials and the Mechanism of Ultrafast Ion Conduction via Dimers*, Springer Theses,
https://doi.org/10.1007/978-981-96-2552-9_3

## 3.1 Introduction

The development of clean energy technologies, such as solid-oxide fuel cells (SOFCs) and proton conducting fuel cells (PCFCs), is an essential path to achieve a carbon–neutral society [1–6]. When considering the cost and stability of SOFCs, it is important to lower the operating temperature, because the high operating temperature causes the degradation due to undesirable chemical reactions and increases their costs [7, 8]. Therefore, a high oxide-ion conductor that operates at low temperature is required. In addition to oxide-ion conductors, the transport and structural properties of proton conductors have been studied because they can exhibit high conductivity at intermediate temperatures due to the low activation energy for conductivity. However, proton conductors used as electrolytes in PCFCs, suffer from poor chemical stability under the $CO_2$ atmosphere and require very high sintering temperatures [9, 10].

To solve the problems of SOFCs and PCFCs, dual-ion conductors where both oxide ion and proton diffuse have been proposed [11–13]. The dual-ion electrolytes have advantages of both oxide-ion and proton conducting electrolytes. First, the simultaneous transport of oxide ion and proton in the electrolyte contributes a lot to the enhancement in total conductivity and an improved cell performance are expected [14]. Second, water will be produced at the cathode and anode, and external gas humidification is not required, which will significantly simplify the fuel cell system [12]. Recently, dual-ion, oxide-ion, and proton conducting hexagonal perovskite-related oxides such as $Ba_7Nb_4MoO_{20}$-based materials have attracted much attention due to high oxide-ion and proton conductivity [10, 13, 15–21]. The bulk conductivities of $Ba_7Nb_4MoO_{20}$ are 2.9 mS cm$^{-1}$ at 500 °C under wet air and 3.0 mS cm$^{-1}$ at 593 °C under dry air [13] but these conductivities are not sufficiently high. Therefore, in this work, we investigate the electrical conductivity of $Ba_7Nb_{4-x}Mo_{1+x}O_{20+x/2}$ materials with Mo excess compositions, which can lead to a larger amount of excess oxygen atoms and higher conductivities. Here we report that $Ba_7Nb_{3.8}Mo_{1.2}O_{20.1}$ exhibits bulk conductivities 11 mS cm$^{-1}$ at 500 °C under wet air (3.6 times higher than $Ba_7Nb_4MoO_{20}$) and 10 mS cm$^{-1}$ at 593 °C under dry air and 175 and 5 times higher bulk conductivity at 306 °C in dry air compared with yttria-stabilized zirconia (YSZ) and $Ba_7Nb_4MoO_{20}$, respectively.

Improved ion conductivity requires deep understanding of the atomic-level mechanisms of ion diffusion and incorporation of oxygen atoms due to the excess oxygen and hydration. The oxide-ion conductivity of $Ba_7Nb_{3.9}Mo_{1.1}O_{20.05}$ is higher than $(Bi_{0.88}Dy_{0.08}W_{0.04})_2O_{3.12}$ at low temperature, which was attributed to the low activation energy for conductivity [15]. However, the origin of the low activation energy has not been elucidated yet. In this work, we resolve the origin of the low activation energy through the interatomic distances. Interstitialcy diffusion mechanism includes a cooperative knock-on motion of interstitial and lattice ions, which has been reported in Ruddlesden-Popper phases [22], melilite-type oxides [23], scheelite-type oxides [24], and hexagonal perovskite derivatives [15]. We have demonstrated the interstitialcy mechanism of oxide-ion diffusion in $Ba_7Nb_4MoO_{20}$-based materials using the neutron scattering length density analysis based on the average structure from neutron diffraction and maximum-entropy method (MEM), but the dynamic local structure in the interstitialcy oxide-ion diffusion is unknown [15, 17, 18]. The ab initio molecular dynamics (AIMD) simulation suggested that high dynamic and rotational flexibility of the variable coordination of a Nb/Mo cation in (Nb/Mo)–O polyhedra is a crucial factor for fast ion conduction in $Ba_7Nb_4MoO_{20}$ [25]. However, the detailed changes in the (Nb/Mo)–O polyhedra during the interstitialcy oxide-ion diffusion has not been studied yet. The oxide-ion diffusion mechanism of $Ba_7Nb_4MoO_{20}$-based materials is an important unresolved issue. Here, we report that the breaking and reforming of $(Nb/Mo)_2O_9$ dimers enable (1) the fast oxide-ion diffusion and (2) the incorporation of excess oxygen atoms in the Mo/Nb substitution and hydration, through the careful examination of the snapshots during the oxide-ion migration through the AIMD simulation of $Ba_7Nb_{3.8}Mo_{1.2}O_{20.1}$. The oxide-ion diffusion mechanism involving the breaking and reforming of dimers has been reported in several materials such as $La_{1-x}Ba_{1+x}GaO_{4-0.5x}$ and $CeNbO_{4+\delta}$ [24, 26]. In this work, we show that $Ba_7Nb_{3.8}Mo_{1.2}O_{20.1}$ exhibits the highest oxide-ion conductivity among these materials, which indicates the unique feature of the present oxide-ion diffusion mechanism in $Ba_7Nb_{3.8}Mo_{1.2}O_{20.1}$. Fop et al. reported proton conduction in the mother material $Ba_7Nb_4MoO_{20}$, [13] which can be ascribed to the proton diffusion in the oxygen deficient $BaO_2$ c′ layer [25]. In this work, we also report higher proton conduction in $Ba_7Nb_{3.8}Mo_{1.2}O_{20.1}$ and its structural origin, where the protons exhibit faster migration in the hexagonal close-packed $BaO_3$ layers. These present findings open new avenues for the development and design of dual-ion-, oxide-ion-, and proton-conducting hexagonal perovskite-related oxides.

## 3.2 Methods

**Sample preparation.** $Ba_7Nb_{4-x}Mo_{1+x}O_{20+x/2}\cdot y\ H_2O$ ($x = 0$, 0.02, 0.04, 0.06, 0.08, 0.1, 0.12, 0.14, 0.16, 0.18, 0.2, 0.22, 0.25, 0.3) samples were synthesized by the solid-state reaction method. Stoichiometric amounts of $BaCO_3$, $Nb_2O_5$, $MoO_3$ (> 99.9%, Kojundo Chemical Co., Ltd.) were mixed and ground as ethanol slurries and dried powders using an agate mortar for 1 h, followed by calcination at 900 °C for 12 h in static air. The obtained samples were ground again for 1 h, pressed uniaxially into pellets at about 150 MPa, and sintered in static air at 1100 °C for 24 h. The sintered pellets were crushed and ground for 1 h into powders. The powders thus obtained were ball-milled using yttria-stabilized zirconia (YSZ) balls (Fritsch, P7). The ball-milled powders were isostatically pressed into pellets at 150 MPa and subsequently sintered in static air at 1200 °C for 12 h. Parts of the pellets sintered at 1200 °C were crushed and ground into powders for characterization. We describe the chemical composition as $Ba_7Nb_{4-x}Mo_{1+x}O_{20+x/2}\cdot y\ H_2O$ for simplicity, although these samples contain small amounts of water. Here, $y$ is the amount of $H_2O$ in $Ba_7Nb_{4-x}Mo_{1+x}O_{20+x/2}\cdot y\ H_2O$, and the $y$ depends on the temperature, humidity, carrier and so on.

**Sample characterization**. The chemical composition of $Ba_7Nb_{3.8}Mo_{1.2}O_{20.1}·y\ H_2O$ was analyzed by ICP-OES. The obtained cation ratio of $Ba_7Nb_{3.8}Mo_{1.2}O_{20.1}·y\ H_2O$ was Ba:Nb:Mo = 6.90(1):3.755(1):1.249(1) ≅ 7:3.8:1.2, which is the same as that of the nominal composition. The phase purities of $Ba_7Nb_{4-x}Mo_{1+x}O_{20+x/2}·y\ H_2O$ ($x$ = 0, 0.02, 0.04, 0.06, 0.08, 0.1, 0.12, 0.14, 0.16, 0.18, 0.2, 0.22, 0.25, and 0.3) were examined by a laboratory-based X-ray diffractometer (Bruker AXS D8 Advance) with Cu $K\alpha$ radiation. X-ray powder diffraction (XRD) data of the mixture of silicon internal standard and $Ba_7Nb_{4-x}Mo_{1+x}O_{20+x/2}·y\ H_2O$ samples ($x$ = 0, 0.04 0.08 0.12 0.16, 0.2, 0.22, and 0.25) were also measured to investigate the lattice parameters by Le Bail analysis using the computer program RIETAN-FP [27]. XPS spectra of $Ba_7Nb_{3.8}Mo_{1.2}O_{20.1}·y\ H_2O$ powder were measured by an X-ray photoelectron spectrometer (JPS 9010, JEOL Ltd.). $Ba_7Nb_{3.8}Mo_{1.2}O_{20.1}·y\ H_2O$ pellets were annealed at 400 and 700 °C for 24 h in $CO_2$ gas flow, and the existing phases in the annealed samples were examined by the laboratory-based X-ray diffractometer.

**Thermogravimetric measurements**. Thermogravimetric (TG) data of $Ba_7Nb_4MoO_{20}·y\ H_2O$ and $Ba_7Nb_{3.8}Mo_{1.2}O_{20.1}·y\ H_2O$ were collected with NETZSCH STA 449 F3. Samples were heated to 900 °C at 10 °C $min^{-1}$ in dry air and held at 900 °C for 1 h. The dry air flow was changed to wet air flow ($P(H_2O)$ = 0.021 atm), and then samples were equilibrium every 100 °C from 900 to 100 °C and 30 °C, with 4 h stabilization time at each temperature. Thermogravimetric and mass spectroscopy (TG-MS) data of $Ba_7Nb_{3.8}Mo_{1.2}O_{20.1}·y\ H_2O$ were collected from 22 to 900 °C with RIGAKU Thermo Mass Photo at the heating rate of 20 °C $min^{-1}$ in He gas flow. The TG-MS measurements were repeated two times to examine the influence of water and carbonates in the sample.

**Electrical conductivity and transport number.** The $Ba_7Nb_{4-x}Mo_{1+x}O_{20+x/2}·y\ H_2O$ pellets ($x$ = 0, 0.02, 0.04, 0.06, 0.08, 0.1, 0.12, 0.14, 0.16, 0.18, 0.2, 0.22, 0.25, and 0.3) (relative density > 95%) were used for measurements of total direct current (DC) electrical conductivities $\sigma_{tot}$, bulk conductivity $\sigma_{bulk}$, and oxide-ion transport numbers $t_{O^{2-}}$. Pt paste and Pt wires were attached to the sintered pellets and heated in static air at 1000 °C for 1 h to remove organics from the Pt paste. The $\sigma_{tot}$ of $Ba_7Nb_{4-x}Mo_{1+x}O_{20+x/2}·y\ H_2O$ were measured by the DC 4-terminal method in static air for $x$ = 0, 0.02, 0.04, 0.06, 0.08, 0.1, 0.12, 0.14, 0.16, 0.18, 0.2, 0.22, 0.25, 0.3 and dry air for $x$ = 0.2 and wet air for $x$ = 0.2 at temperatures from 304 to 907 °C. The $\sigma_{tot}$ of the $Ba_7Nb_{3.8}Mo_{1.2}O_{20.1}·y\ H_2O$ pellet was measured at various oxygen partial pressures $P(O_2)$ at 302 °C, 603 °C, and 908 °C in dry atmospheres. The $P(O_2)$ was controlled using a mixture of $O_2$, $N_2$, and 5% $H_2$ in $N_2$ gas, and monitored by an oxygen sensor. The $t_{O^{2-}}$ of $Ba_7Nb_{3.8}Mo_{1.2}O_{20.1}$ were examined by an oxygen concentration cell at 596, 697, 800, and 903 °C after heating at 1000 °C for 1 h in dry air. One side of the pellet was exposed to flowing dry air and the other side to flowing dry $O_2$ gas (dry air)/(dry $O_2$), mixture gas of dry air and dry $N_2$ (dry air)/(dry 10% air + dry 90% $N_2$), and dry 5% $H_2$ in $N_2$ gas (dry air)/(5% $H_2$ in $N_2$). The electromotive force $E$ of the concentration cell was recorded with a Keithley model 617 electrometer. The following Nernst equation was used to estimate the $t_{O^{2-}}$:

$$E = t_{O^{2-}} \frac{RT}{4F} \ln\left(\frac{p(O_2)}{p^0(O_2)}\right)$$

where $F$ is the Faraday constant, $R$ is the gas constant, $t$ is the absolute temperature, $P(O_2)$ is the oxygen partial pressure of the gas of $O_2$, $N_2$, 5% $H_2$ in $N_2$, and $p^0(O_2)$ (= 0.21 atm) is the oxygen partial pressure of dry air. The bulk, grain boundary, and electrode conductivities of $Ba_7Nb_{3.8}Mo_{1.2}O_{20.1}$ and $Ba_7Nb_{3.8}Mo_{1.2}O_{20.1} \cdot y\ H_2O$ were evaluated as a function of temperature from 593 to 31 °C on cooling after heating up to 900 °C by AC impedance spectroscopy in flowing dry and wet air, respectively, (100 mL min$^{-1}$) using a pellet with relative density higher than 98% and Pt electrodes. Impedance spectra were recorded using a Solartron 1260 impedance analyzer in the frequency range of 10 MHz – 1 Hz at an applied alternating voltage of 100 mV. The *Lin-KK* tool software was employed to perform the KK transformation on the collected impedance data [28].

**Neutron-diffraction measurements and data analyses.** Neutron-diffraction measurements of $Ba_7Nb_4MoO_{20}$ pellets were performed at 800 °C in vacuum using the super-high-resolution time-of-flight (TOF) neutron diffractometer (SuperHRPD) installed Materials and Life Science Experimental Facility of J-PARC, Japan [29, 30]. Neutron-diffraction data of $Ba_7Nb_{3.8}Mo_{1.2}O_{20.1} \cdot y\ H_2O$ pellets were measured at 21 °C with SPICA diffractometer at the MLF in J-PARC [31]. Neutron-diffraction data of $Ba_7Nb_{3.8}Mo_{1.2}O_{20.1}$ pellets were measured also at 800 °C in vacuum with the Echidna instrument at the Australian Nuclear Science and Technology Organization (ANSTO) (wavelength: $\lambda$ = 1.622425(10) Å, step interval: 0.125° in 2 $\theta$) [32].

To prepare a dry $Ba_7Nb_{3.8}Mo_{1.2}O_{20.1}$ sample, $Ba_7Nb_{3.8}Mo_{1.2}O_{20.1}$ pellets were annealed at 800 °C for 2h in a vacuum sealed quartz tube. To synthesize a wet $Ba_7Nb_{3.8}Mo_{1.2}O_{20.1}$ sample, the $Ba_7Nb_{3.8}Mo_{1.2}O_{20.1}$ pellets were annealed at 800 °C for 1h in $D_2O$/air (water vapor pressure $P(D_2O)$ = 0.021 atm) and then cooled at the cooling rate of 5 °C/min in the $D_2O$/air. The dry and wet $Ba_7Nb_{3.8}Mo_{1.2}O_{20.1}$ samples were put into vanadium holders under He gas flow, and their neutron diffraction data were measured at 5K by the HERMES diffractomer [33] installed in the JRR-3 reactor (wavelength: $\lambda$ = 1.34171(5) Å, step interval: 0.05° in 2 $\theta$).

The data collected at SPICA and SuperHRPD were analyzed by the Rietveld method using the Z-Rietveld program [34, 35], HERMES and Echidna were analyzed by the Rietveld method using the RIETAN-FP [27], respectively. The neutron scattering length density distribution was investigated using the maximum entropy method (MEM). The MEM analysis was carried out with Z-MEM software [36] and Dysnomia [37] using the structure factors obtained from the Rietveld refinement of the neutron diffraction data. The MEM calculations were performed with the unit cell divided into 60 × 60 × 168 pixels.

The bond-valence-based energy (BVE) calculations were carried out using the computer program *softBV* [38, 39]. The Madelung energies were calculated by *VESTA 3* [40]. The crystal structures, neutron scattering length density distributions, and BVE landscapes were visualized using the VESTA program [40].

**Ab initio molecular dynamics (AIMD) simulations.** AIMD simulations were carried out the Vienna ab initio simulation package (VASP) at a constant temperature within the canonical ensemble (NVT) using a Nosé thermostat. The system was heated to the target temperature (800 and 1200 °C) at a rate of 1 K/fs, and further isothermal MD simulations were carried out for 60–100 ps with a time step of 2 fs. Due to the large size of the supercells (~ 17.6 × 17.6 × 16.5 Å), integration in reciprocal space was performed only at the Γ-point.

## 3.3 Result and Discussion

### 3.3.1 *Synthesis and Characterization of* $Ba_7Nb_{4-x}Mo_{1+x}O_{20+x/2}\cdot y\ H_2O$

Samples of $Ba_7Nb_4MoO_{20}\cdot y\ H_2O$, $Ba_7Nb_{3.9}Mo_{1.1}O_{20.05}\cdot y\ H_2O$ and new materials $Ba_7Nb_{4-x}Mo_{1+x}O_{20+x/2}\cdot y\ H_2O$ ($x$ = 0.02, 0.04, 0.06, 0.08, 0.12, 0.14, 0.16, 0.18, 0.2, 0.22, 0.25, and 0.3) were synthesized by the solid-state-reaction method [15]. Here the $y$ is the water content in $Ba_7Nb_{4-x}Mo_{1+x}O_{20+x/2}\cdot y\ H_2O$ and the water is mainly incorporated as hydroxide ions in bulk $Ba_7Nb_{4-x}Mo_{1+x}O_{20+x/2+y}H_{2y}$ [= $Ba_7Nb_{4-x}Mo_{1+x}O_{20+x/2-y}(OH)_{2y}$] as shown later for the $x = 0.2$ composition. Figure 3.1 shows the X-ray powder diffraction (XRD) patterns of $Ba_7Nb_{4-x}Mo_{1+x}O_{20+x/2}\cdot y\ H_2O$ at room temperature (RT). The reflections were indexed based on the primitive hexagonal *hP* lattice, identifying $Ba_7Nb_{4-x}Mo_{1+x}O_{20+x/2}\cdot y\ H_2O$ ($0 \leq x \leq 0.2$) to 7H polytype (crystal system: trigonal, space group: $P\overline{3}m1$ [41]. Meanwhile, $Ba_7Nb_{4-x}Mo_{1+x}O_{20+x/2}\cdot y\ H_2O$ ($0.22 \leq x \leq 0.3$) was identified to be main $P\overline{3}m1$ hexagonal phase, in addition to small amounts of $Ba_3MoNbO_{8.5}$. The lattice volume of $Ba_7Nb_{4-x}Mo_{1+x}O_{20+x/2}\cdot y\ H_2O$ decreases linearly with an increase of excess Mo content $x$ up to $x = 0.2$ (Fig. 3.2; Table 3.1) indicating the formation of solid solutions $Ba_7Nb_{4-x}Mo_{1+x}O_{20+x/2}\cdot y\ H_2O$ ($0 < x \leq 0.2$). The decrease in the lattice volume is ascribed to the smaller ionic radius of $Mo^{6+}$ (0.41 Å in fourfold coordination) in comparison with that of $Nb^{5+}$ (0.48 Å in fourfold coordination) [42]. The Mo solubility limit in $Ba_7Nb_{4-x}Mo_{1+x}O_{20+x/2}\cdot y\ H_2O$ was estimated to be $x = 0.196(10)$ from the extrapolation of a straight-line fit to the lattice volumes.

Total direct current (DC) electrical conductivities $\sigma_{tot}$ of $Ba_7Nb_{4-x}Mo_{1+x}O_{20+x/2}\cdot y\ H_2O$ were measured by the four-probe method. The $\sigma_{tot}$ increases with temperature in static air (Fig. 3.3). At a constant temperature from 304 to 907 °C, the $\sigma_{tot}$ increases with an increase of excess Mo content $x$ (excess oxygen amount $x/2$) and reaches a maximum at $x = 0.2$, while $\sigma_{tot}$ decreases with $x$ beyond that (Fig. 3.4) probably due to the increase of the amount of impurity phase. For example, the $\sigma_{tot}$ at 300 °C in static air for $x = 0.2$ is 16 times higher than that for the mother material ($x = 0.0$). $Ba_7Nb_{3.8}Mo_{1.2}O_{20.1}\cdot y\ H_2O$ also exhibits higher $\sigma_{tot}$ in static air compared with other 7H polytypes such as $Ba_7Nb_{3.8}Cr_{0.2}MoO_{20.1}\cdot y\ H_2O$ [17] and

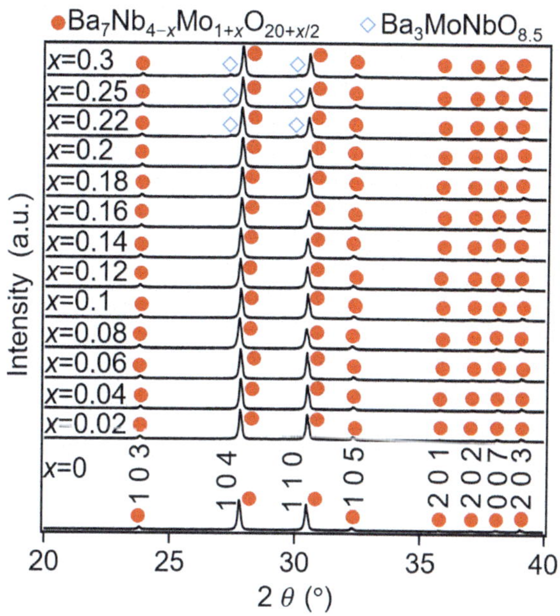


**Fig. 3.1** X-ray powder diffraction patterns of $Ba_7Nb_{4-x}Mo_{1+x}O_{20+x/2}\cdot y\ H_2O$ ($x = 0$, 0.02, 0.04, 0.06, 0.08, 0.1, 0.12, 0.14, 0.16, 0.18, 0.2, 0.22, 0.25, 0.3). Closed red circles and open blue rhombuses denote peaks of hexagonal $P\overline{3}m1$ 7H-polytype $Ba_7Nb_{4-x}Mo_{1+x}O_{20+x/2}\cdot y\ H_2O$ and rhombohedral $Ba_3MoNbO_{8.5}$, respectively. Copyright© 2023 Authors

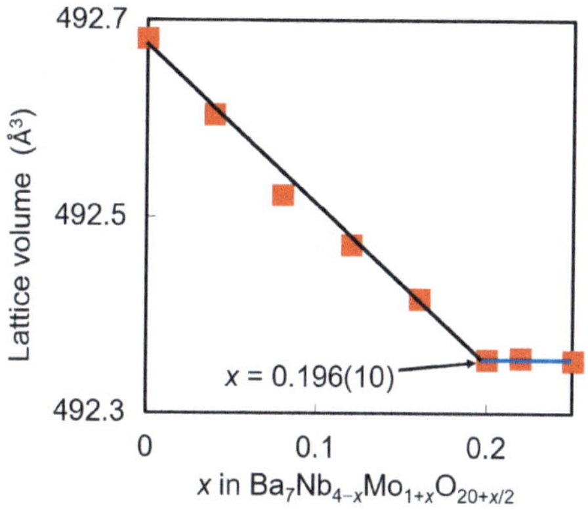


**Fig. 3.2** Lattice volume of $Ba_7Nb_{4-x}Mo_{1+x}O_{20+x/2}\cdot y\ H_2O$ ($x = 0$, 0.04, 0.08, 0.12, 0.16, 0.2, 0.22, 0.25) versus $x$ at RT. The arrow shows the solubility limit $x = 0.196(10)$ in $Ba_7Nb_{4-x}Mo_{1+x}O_{20+x/2}\cdot y\ H_2O$. For $x$ values higher than 0.2, the lattice volume almost remains constant, and the XRD peak intensities of impurity phases increase with $x$. The Mo solubility limit in $Ba_7Nb_{4-x}Mo_{1+x}O_{20+x/2}\cdot y\ H_2O$ is estimated to be $x = 0.196(10)$ from the extrapolation of a straight-line fit to the lattice volumes. Copyright© 2023 Authors

$Ba_7Nb_{3.85}W_{0.15}MoO_{20.075}\cdot y\ H_2O$ [18] (Fig. 3.5). Further studies are focused on the $x = 0.2$ composition, as it has the highest $\sigma_{tot}$ in the $Ba_7Nb_{4-x}Mo_{1+x}O_{20+x/2}\cdot y\ H_2O$ samples ($0 \leq x \leq 0.3$). Inductively coupled plasma optical emission spectroscopy (ICP-OES) for $Ba_7Nb_{3.8}Mo_{1.2}O_{20.1}\cdot y\ H_2O$ ($x = 0.2$) showed that the atomic ratio of

**Table 3.1** Lattice parameters and lattice volumes of $Ba_7Nb_{4-x}Mo_{1+x}O_{20+x/2}\cdot y\ H_2O$ ($x = 0$, 0.04, 0.08, 0.12, 0.16, 0.2, 0.22, 0.25) versus the excess Mo content $x$ at RT

| Composition $x$ | Lattice parameters (Å) | | Lattice volume (Å$^3$) |
|---|---|---|---|
| | $a$ | $c$ | |
| 0 | 5.8659(3) | 16.5334(5) | 492.68(6) |
| 0.04 | 5.8645(3) | 16.5387(4) | 492.60(6) |
| 0.08 | 5.8645(3) | 16.5359(4) | 492.52(4) |
| 0.12 | 5.8643(3) | 16.5356(4) | 492.47(4) |
| 0.16 | 5.8647(4) | 16.5316(6) | 492.41(7) |
| 0.2 | 5.8646(3) | 16.5299(4) | 492.35(5) |
| 0.22 | 5.8641(3) | 16.5328(4) | 492.35(5) |
| 0.25 | 5.8646(4) | 16.5302(5) | 492.35(6) |

The numbers in the parentheses are the estimated standard deviation of the last digit

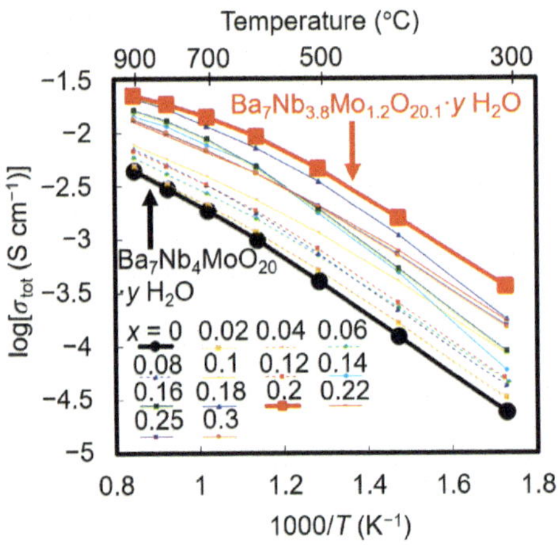


**Fig. 3.3** Temperature dependence of total DC electrical conductivities $\sigma_{tot}$ of $Ba_7Nb_{4-x}Mo_{1+x}O_{20+x/2}\cdot y\ H_2O$ ($x = 0$, 0.02, 0.04, 0.06, 0.08, 0.1, 0.12, 0.14, 0.16, 0.18, 0.2, 0.22, 0.25, and 0.3) in static air from 304 to 907 °C. Copyright© 2023 Authors

the nominal composition Ba:Nb:Mo = 7:3.8:1.2. X-ray photoelectron spectroscopy (XPS) spectra of the $x = 0.2$ composition indicated that Mo and Nb adopt + 6 and + 5 oxidation states, respectively (Fig. 3.6). Therefore, the chemical composition of the $x = 0.2$ sample is given by $(Ba^{2+})_7(Nb^{5+})_{3.8}(Mo^{6+})_{1.2}(O^{2-})_{20.1}\cdot y\ H_2O$.

### 3.3.2 High Oxide-Ion Conduction, High Chemical and Electrical Stability of $Ba_7Nb_{3.8}Mo_{1.2}O_{20.1}$

Figure 3.7a shows the oxygen partial pressure $P(O_2)$ dependence of $\sigma_{tot}$ for $Ba_7Nb_{3.8}Mo_{1.2}O_{20.1}$ at 302, 603, and 908 °C in dry atmospheres. The $\sigma_{tot}$ is almost independent of $P(O_2)$, in the $P(O_2)$ ranges from 1.0 to 8.9 × $10^{-25}$ atm

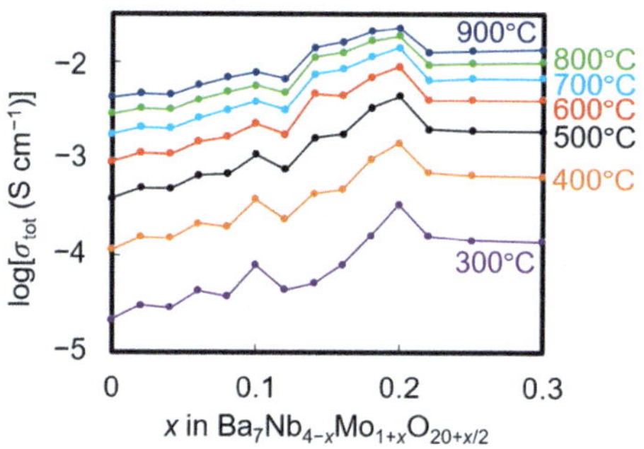


**Fig. 3.4** Excess Mo content $x$ dependence of total DC electrical conductivities $\sigma_{tot}$ of $Ba_7Nb_{4-x}Mo_{1+x}O_{20+x/2} \cdot y\ H_2O$ in static air at constant temperatures from 300 to 900 °C. The $\sigma_{tot}$ was calculated by interpolating the data in Fig. 3.3. The $\sigma_{tot}$ at 300 °C was estimated by extrapolating the data in Fig. 3.3. Copyright© 2023 Authors

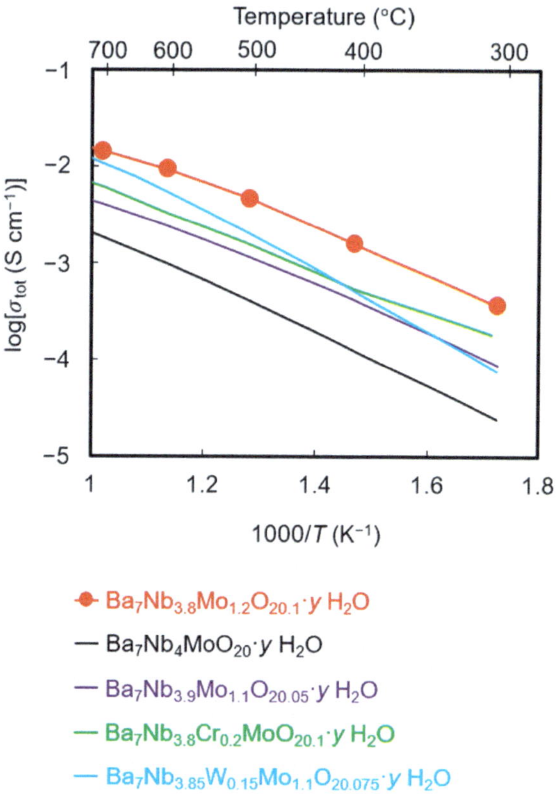


**Fig. 3.5** Comparison of total DC electrical conductivity $\sigma_{tot}$ in static air of $Ba_7Nb_{3.8}Mo_{1.2}O_{20.1} \cdot y\ H_2O$ (this work) and other $Ba_7Nb_4MoO_{20}$-based oxides: $Ba_7Nb_4MoO_{20} \cdot y\ H_2O$ (this work), $Ba_7Nb_{3.9}Mo_{1.1}O_{20.05} \cdot y\ H_2O$ (this work), $Ba_7Nb_{3.8}Cr_{0.2}MoO_{20.1} \cdot y\ H_2O$ [17], and $Ba_7Nb_{3.85}W_{0.15}MoO_{20.075} \cdot y\ H_2O$ [18]. Copyright© 2023 Authors

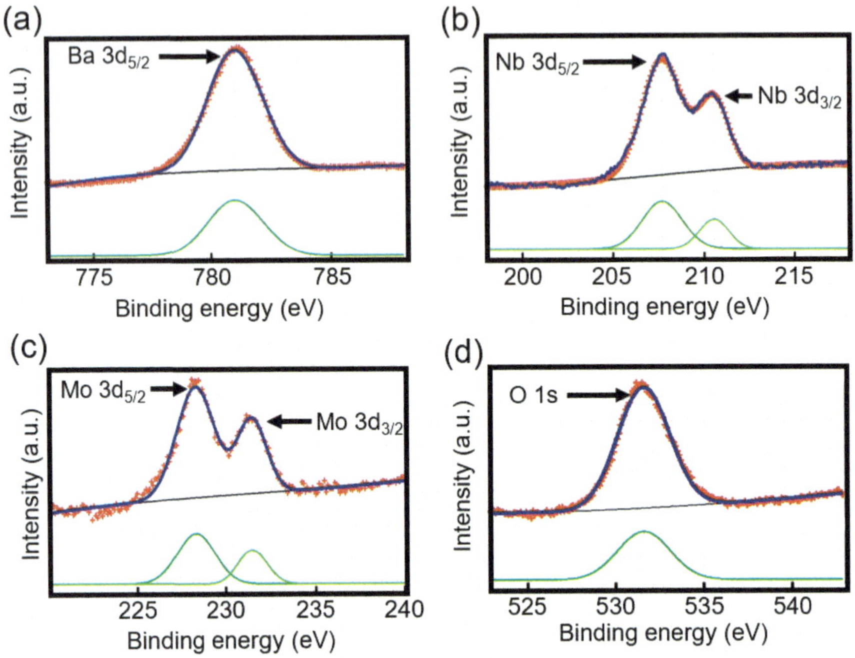


**Fig. 3.6** X-ray photoelectron spectroscopy (XPS) spectra fitting of (a) Ba 3d, (b) Nb 3d, (c) Mo 3d, and (d) O 1s states of $Ba_7Nb_{3.8}Mo_{1.2}O_{20.1} \cdot y\ H_2O$ at RT. The observed intensities, fitting intensities, background, calculated components are shown by red cross marks, dark blue lines, black lines, and (light) green lines, respectively. By the fitting, we obtained the following binding energies, which agree well with the reported values described in the parentheses. Ba $3d_{5/2}$: 781 eV (780 eV) [43], Nb $3d_{5/2}$: 207.7 eV (207.2 eV) [44], Nb $3d_{3/2}$: 210.5 eV (210 eV) [44], Mo $3d_{5/2}$: 233.3 eV (232.3–232.6 eV) [45], and Mo $3d_{3/2}$: 236.5 eV (235.4–235.7 eV) [45]. Each obtained value agrees with the reported one, thus, the valences of Ba, Nb and Mo are + 2, + 5, and + 6, respectively. XPS spectra of the $x = 0.2$ composition indicate that Mo and Nb adopt + 6 and + 5 oxidation states, respectively. The XPS results, combined with the charge neutrality condition, led us to conclude that the chemical composition of the $x = 0.2$ sample is given by $(Ba^{2+})_7(Nb^{5+})_{3.8}(Mo^{6+})_{1.2}(O^{2-})_{20.1} \cdot y\ H_2O$ at RT. Copyright© 2023 Authors

at 302 °C and 1.0 to 5.4 × $10^{-25}$ atm at 603 °C, indicating the extremely wide electrolyte domain of $Ba_7Nb_{3.8}Mo_{1.2}O_{20.1}$. At 908 °C, the electrolyte domain is slightly narrowed. The slope in the low $P(O_2)$ range between 1.2 × $10^{-24}$ and 5.6 × $10^{-21}$ atm at 908 °C has a negative value of − 0.0276(5), implying *n*-type electronic conduction probably caused by partial reduction of $Nb^{5+}$ and/or $Mo^{6+}$ cations. Compared with $Ba_7Nb_4MoO_{20}$, [13] the present $Ba_7Nb_{3.8}Mo_{1.2}O_{20.1}$ sample shows a wide electrolyte domain, similar to $Ba_7Nb_{3.9}Mo_{1.1}O_{20.05}$, [15] $Ba_7Nb_{3.8}Cr_{0.2}MoO_{20.1}$, [17] and $Ba_7Nb_{3.85}W_{0.15}MoO_{20.075}$, [18] demonstrating the high chemical and electrical stability of $Ba_7Nb_{3.8}Mo_{1.2}O_{20.1}$. Oxide-ion transport number $t_{O^{2-}}$ of $Ba_7Nb_{3.8}Mo_{1.2}O_{20.1}$ was evaluated using an oxygen-concentration cell (Fig. 3.7(b)). The obtained $t_{O^{2-}}$ are 0.95–0.98 between 697 and 903 °C in (dry air)/(dry 10% air in $N_2$), 0.92 − 1.0 between 697 and 903 °C in (dry air)/

(dry $O_2$), and 1.0 at 596 °C in (dry air)/(dry 5% $H_2$ in $N_2$), demonstrating that $Ba_7Nb_{3.8}Mo_{1.2}O_{20.1}$ is an oxide-ion conductor with negligible electronic contribution in dry conditions. It should be noted that $Ba_7Nb_{3.8}Mo_{1.2}O_{20.1}$ exhibits enhanced phase stability under $CO_2$ flow at high temperatures, compared with $Ba_7Nb_4MoO_{20}$. XRD patterns of $Ba_7Nb_{3.8}Mo_{1.2}O_{20.1}$ samples before and after annealing for 24 h under $CO_2$ flow at 400 °C were in agreement with each other, showing no phase changes. After the annealing for 24 h under $CO_2$ flow at 700 °C, a small amount of impurity phases appeared in $Ba_7Nb_{3.8}Mo_{1.2}O_{20.1}$ (maximum height of the impurity XRD peaks was only 2.6% of the strongest maximum XRD peak height of the hexagonal $Ba_7Nb_{3.8}Mo_{1.2}O_{20.1}$), which might be lower than that (6.0%) in $Ba_7Nb_4MoO_{20}$ [13] (Fig. 3.8).

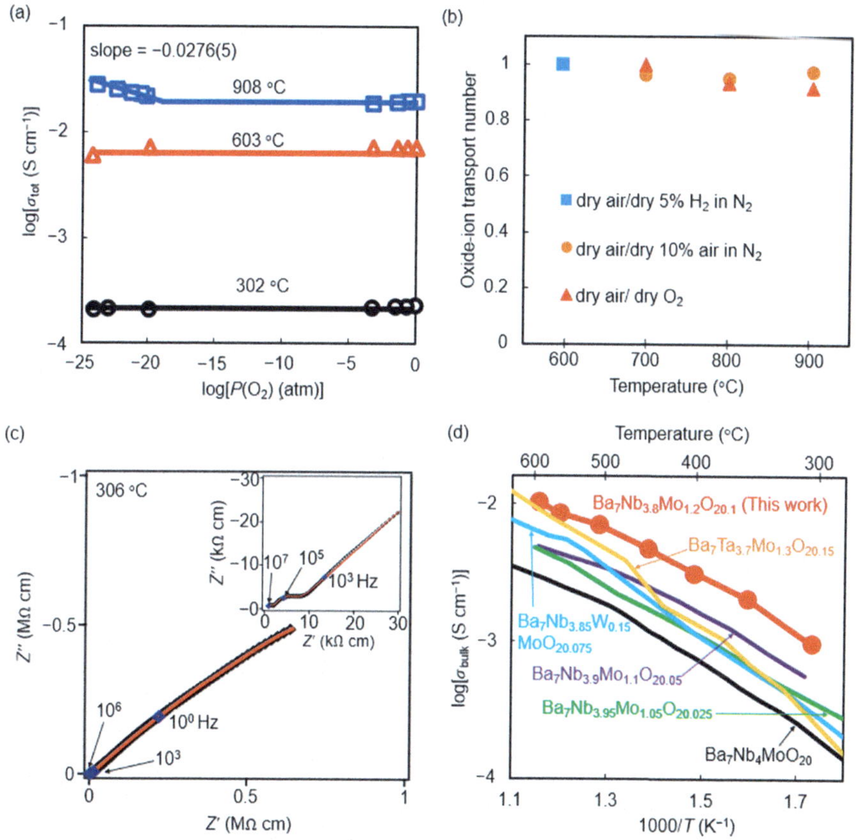


**Fig. 3.7** High oxide-ion conductivity and stability of $Ba_7Nb_{3.8}Mo_{1.2}O_{20.1}$. **a** Partial oxygen pressure $P(O_2)$ dependencies of the $\sigma_{tot}$ for $Ba_7Nb_{3.8}Mo_{1.2}O_{20.1}$ at 302, 603, and 908 °C. **b** Oxide-ion transport numbers $t_{O^{2-}}$ of $Ba_7Nb_{3.8}Mo_{1.2}O_{20.1}$ from 596 to 903 °C. **c** Complex impedance plots of $Ba_7Nb_{3.8}Mo_{1.2}O_{20.1}$ recorded at 306 °C in dry air. **d** Comparison of bulk conductivity of $Ba_7Nb_{3.8}Mo_{1.2}O_{20.1}$ and other $Ba_7Nb_4MoO_{20}$-based oxides [13, 15–18] in dry air. Copyright© 2023 Authors

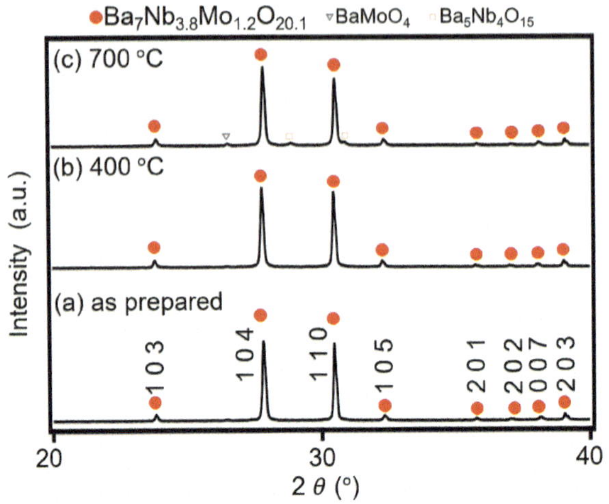


**Fig. 3.8** XRD patterns of $Ba_7Nb_{3.8}Mo_{1.2}O_{20.1}\cdot y\ H_2O$ **a** before and **b**, **c** after annealing under flowing $CO_2$ gas at different temperatures for 24 h. Reflection indices *hkl* in panel (a) are based on the hexagonal setting. The phase stability under $CO_2$ gas flow of $Ba_7Nb_{3.8}Mo_{1.2}O_{20.1}\cdot y\ H_2O$ is higher than that of $Ba_7Nb_4MoO_{20}\cdot y\ H_2O$ [13]. Height ratio of intensity: $(Ba_5Nb_4O_{15})$/ $(Ba_7Nb_{3.8}Mo_{1.2}O_{20.1}\cdot y\ H_2O) = 2.6\%$. Copyright© 2023 Authors

Typical impedance spectra of $Ba_7Nb_{3.8}Mo_{1.2}O_{20.1}$ in dry air and $Ba_7Nb_{3.8}Mo_{1.2}O_{20.1}\cdot y\ H_2O$ in wet air are shown in Figs. 3.7c, 3.9, and 3.10. The bulk conductivity $\sigma_{bulk}$ and grain-boundary conductivity $\sigma_{gb}$ were obtained by equivalent circuit (EC) analysis (Fig. 3.11). The $\sigma_{bulk}$ in dry air of $Ba_7Nb_{3.8}Mo_{1.2}O_{20.1}$ is higher than those of $Ba_7Nb_4MoO_{20}$, [13] $Ba_7Nb_{3.95}Mo_{1.05}O_{20.025}$, [15] $Ba_7Nb_{3.9}Mo_{1.1}O_{20.05}$, [15] $Ba_7Ta_{3.7}Mo_{1.3}O_{20.15}$, [16] and $Ba_7Nb_{3.85}W_{0.15}MoO_{20.075}$ [18] in the whole temperature range between 306 and 593 °C (Fig. 3.7d). $\sigma_{bulk}$ of $Ba_7Nb_{3.8}Mo_{1.2}O_{20.1}$ is 10 mS cm$^{-1}$ at 593 °C in dry air. $Ba_7Nb_{3.8}Mo_{1.2}O_{20.1}$ exhibits extremely high $\sigma_{bulk}$ at lower temperatures due to low activation energy ($E_a$ = 0.25(4)–0.41(4) eV in Fig. 3.12). At 306 °C, the $\sigma_{bulk}$ in the dry air of $Ba_7Nb_{3.8}Mo_{1.2}O_{20.1}$ 1.1 mS cm$^{-1}$ is 5.0 times higher than that of $Ba_7Nb_4MoO_{20}$ [13] and even higher than that of stabilized $Bi_2O_3$, [46] $Ba_7Nb_{3.9}Mo_{1.1}O_{20.05}$, [13] and $Ba_7Ta_{3.7}Mo_{1.3}O_{20.15}$ [16]. $Ba_7Nb_{3.8}Mo_{1.2}O_{20.1}$ exhibits much higher $\sigma_{bulk}$ at 306 °C in dry air compared with those of representative oxide-ion conductors (175 times higher in comparison with $(ZrO_2)_{0.92}(Y_2O_3)_{0.08}$ (8YSZ) [47], 36 times higher than $La_{0.8}Sr_{0.2}Ga_{0.83}Mg_{0.17}O_{2.815}$ (LSGM) [48], and 9 times higher compared to $Ce_{0.9}Gd_{0.1}O_{1.95}$; [49] Fig. 3.13a). The $\sigma_{tot}$ of the $x = 0.2$ composition $Ba_7Nb_{3.8}Mo_{1.2}O_{20.1}\cdot y\ H_2O$ is 22 mS cm$^{-1}$ at 912 °C (Fig. 3.14a).

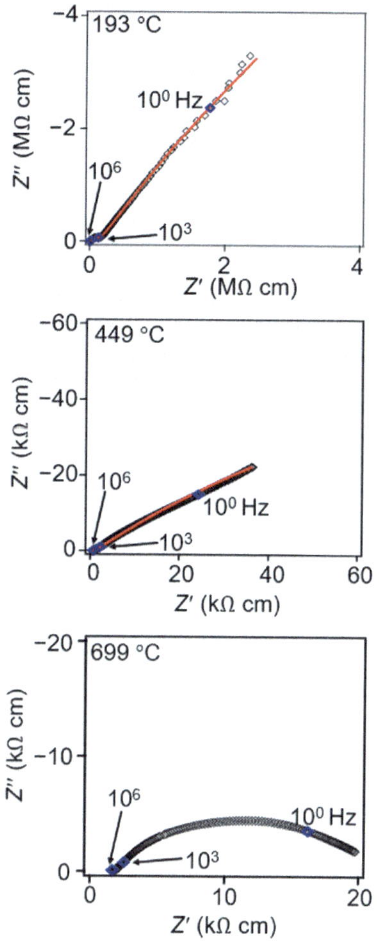


**Fig. 3.9** Complex impedance plots of $Ba_7Nb_{3.8}Mo_{1.2}O_{20.1}$ in dry air from 193 to 699 °C. Black rhombuses and red lines represent measured data and equivalent circuit fittings, respectively. The Warburg electrode responses were observed in the low-frequency range from 193 to 449 °C, and the electrode response collapsed to a semicircular arc at 699 °C, which has been observed in typical oxide-ion conductors such as $Ba_7Nb_4MoO_{20}$, $Ba_7Nb_{3.9}Mo_{1.1}O_{20.05}$ and $Bi_{0.85}Ca_{0.15}VO_{3.925}$ [13, 15, 50]. The small semicircle can be assigned to the bulk response, because the obtained bulk capacitance $C_b$ values are 7–8 pF cm$^{-1}$. The obtained grain-boundary capacitance $C_{gb}$ values of 0.5–7 nF cm$^{-1}$ are characteristic of a grain boundary response. The obtained electrode capacitance $C_{el}$ values of 20–100 μF cm$^{-1}$ are close to the capacitance values of 50–70 μF cm$^{-1}$ in $Ba_7Nb_4MoO_{20}$ at low frequency (510 °C) [13]. Copyright© 2023 Authors

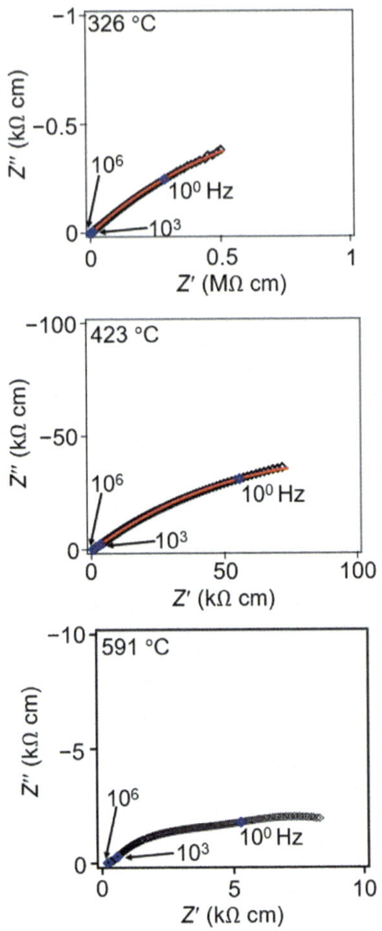


**Fig. 3.10** Complex impedance plots of $Ba_7Nb_{3.8}Mo_{1.2}O_{20.1}{\cdot}y\ H_2O$ in wet air from 326 °C to 591 °C. Black rhombuses and red lines represent measured data and equivalent circuit fittings, respectively. With an increase in temperature, the Warburg electrode response arc collapses, which was also observed in the oxide-ion and proton conductors such as $Ba_7Nb_4MoO_{20}$ and $Sr_3V_2O_8$ [13, 51]. The obtained grain-boundary capacitance $C_{gb}$ values of 0.2–3 nF $cm^{-1}$ are characteristic of a grain-boundary response. The obtained electrode capacitance $C_{el}$ values of 0.01–0.04 μF $cm^{-1}$ are characteristic of a electrode response. Copyright© 2023 Authors

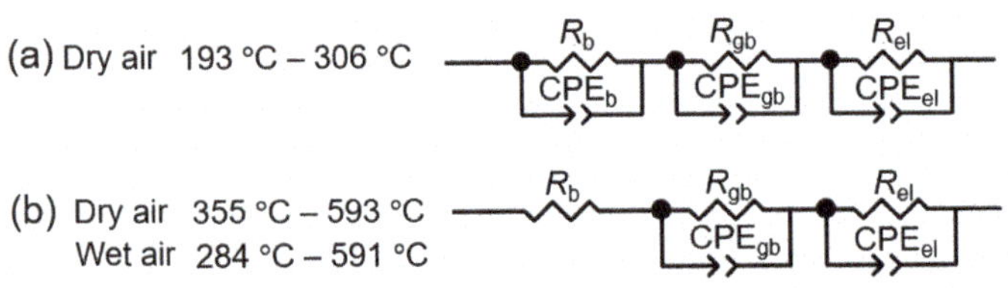


**Fig. 3.11** **a** Equivalent circuits used to model the impedance spectra of $Ba_7Nb_{3.8}Mo_{1.2}O_{20.1}$ from 193 to 306 °C in dry air. **b** Equivalent circuit used to model the impedance spectra of $Ba_7Nb_{3.8}Mo_{1.2}O_{20.1}$ from 355 °C to 593 °C in dry air and $Ba_7Nb_{3.8}Mo_{1.2}O_{20.1}{\cdot}y\ H_2O$ from 284 °C to 591 °C in wet air. *R* and CPE denote a resistor and a constant phase element, respectively. The subscripts b, gb, and el represent bulk, grain boundary, and electrode, respectively. Copyright© 2023 Authors

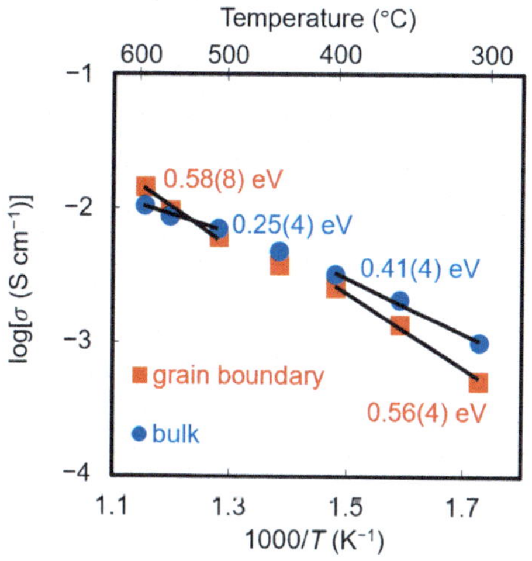


**Fig. 3.12** Arrhenius plots of bulk conductivity $\sigma_{bulk}$ (blue circles) and grain-boundary conductivity $\sigma_{gb}$ (red squares) of $Ba_7Nb_{3.8}Mo_{1.2}O_{20.1}$ from 306 to 593 °C in dry air. Red and blue numbers denote the values of activation energies for $\sigma_{bulk}$ and $\sigma_{gb}$, respectively. Black solid line stands for the linear fit. The activation energy $E_a$ for $\sigma_{gb}$ is higher than that for $\sigma_{bulk}$: $E_a$ for $\sigma_{gb}$ 0.56(4) eV > $E_a$ for $\sigma_{bulk}$ 0.41(4) eV from 306 to 402 °C. $E_a$ for $\sigma_{gb}$ 0.58(8) eV > $E_a$ for $\sigma_{bulk}$ 0.25(4) eV from 509 to 593 °C. Lower $E_a$ at higher temperatures were also reported in $Ba_7Nb_{4-x}Mo_{1+x}O_{20+x/2}$ ($x = 0, 0.1$) [13, 15]. Copyright© 2023 Authors

### 3.3.3 *High Dual-Ion and Proton Conduction of $Ba_7Nb_{3.8}Mo_{1.2}O_{20.1}$·y $H_2O$*

The total DC electrical conductivity $\sigma_{DC}$ of $Ba_7Nb_{3.8}Mo_{1.2}O_{20.1}\cdot y\ H_2O$ was measured also in wet air (water vapor pressure $P(H_2O) = 0.021$ atm) to examine its dual-ion proton and oxide-ion conductivity and proton conductivity. The $\sigma_{DC}$ in wet air $\sigma^{DC}_{wet}$ is higher than that in dry air (Fig. 3.14a), indicating proton conduction. Proton conductivity was estimated by an equation $\sigma^{DC}_{H^+} = \sigma^{DC}_{wet} - \sigma^{DC}_{dry}$ (Fig. 3.15a). The $\sigma^{DC}_{H^+}$ of $Ba_7Nb_{3.8}Mo_{1.2}O_{20.1}\cdot y\ H_2O$ is 2.4 mS cm$^{-1}$ at 700 °C. At 500 °C, the $\sigma^{DC}_{H^+}$ of $Ba_7Nb_{3.8}Mo_{1.2}O_{20.1}\cdot y\ H_2O$ was 82 times higher than $\sigma^{DC}_{H^+}$ of $Ba_4Nb_4MoO_{20}\cdot y\ H_2O$ [18]. The DC electrical conductivity of $Ba_7Nb_{4-x}Mo_{1+x}O_{20+x/2}\cdot y\ H_2O$ ($x = 0$, 0.1, 0.2) in wet air ($H_2O$ partial pressure, $P(H_2O) = 0.021$ atm) $\sigma^{DC}_{wet}$ increases with increasing of excess Mo content $x$ (Fig. 3.14a). The $\sigma^{DC}_{wet}$ of $Ba_7Nb_{3.8}Mo_{1.2}O_{20.1}\cdot y\ H_2O$ is higher than that of $Ba_7Nb_4MoO_{20}\cdot y\ H_2O$ [18] (13 and 6.2 times higher at 400 and 800 °C, respectively) and 24 mS cm$^{-1}$ at 888 °C (Fig. 3.14a). The bulk (dual-ion proton and oxide-ion) conductivity in wet air $\sigma^{bulk}_{wet}$ of $Ba_7Nb_{3.8}Mo_{1.2}O_{20.1}\cdot y\ H_2O$ is also higher than $\sigma^{bulk}_{wet}$ of $Ba_4Nb_4MoO_{20}\cdot y\ H_2O$ [13] (3.6 times higher at 500 °C) and 11 mS cm$^{-1}$ at 537 °C (Fig. 3.15b). At 537 °C, $Ba_7Nb_{3.8}Mo_{1.2}O_{20.1}\cdot 0.11\ H_2O$ exhibits much higher $\sigma^{bulk}_{wet}$ compared with representative proton conductors (73 times higher in comparison with $La_{0.8}Ba_{1.2}GaO_{3.9}$, [61] 74 times higher than $La_{26}O_{27}(BO_3)_8$, [62] 435 times higher than $La_{5.4}MoO_{11.1}$, [63] 793 times higher than $La_{0.99}Ca_{0.01}VO_4$, [64]

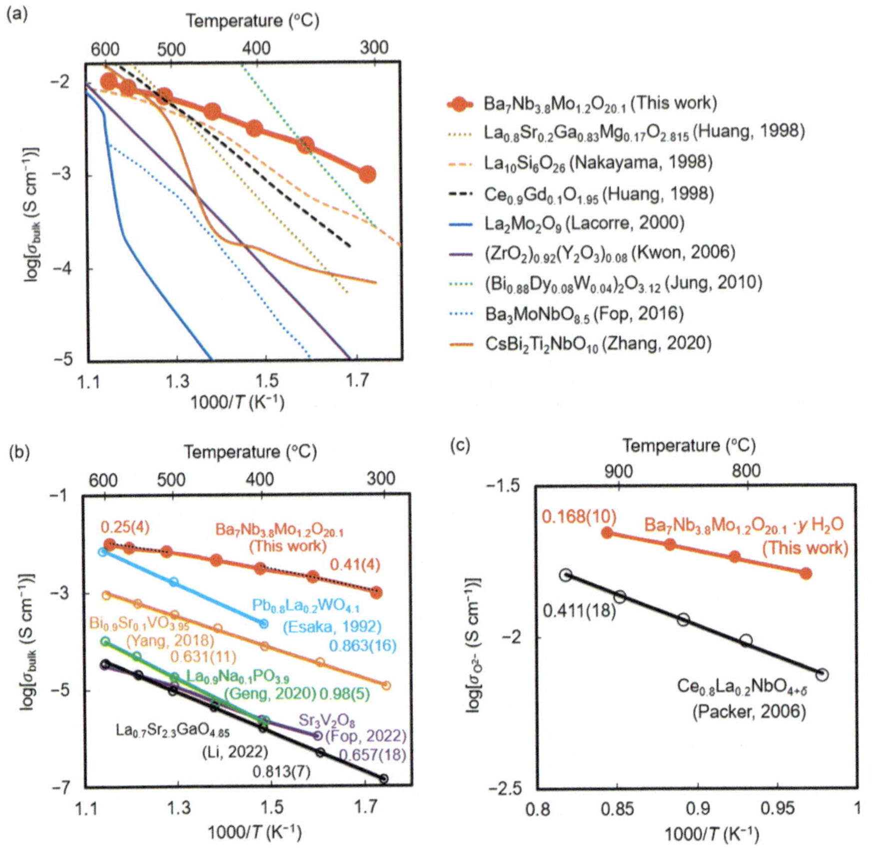


**Fig. 3.13** Comparison of conductivity of $Ba_7Nb_{3.8}Mo_{1.2}O_{20.1}$ and other representative oxide-ion conductors. **a** $\sigma_{bulk}$ of $(Bi_{0.88}Dy_{0.08}W_{0.04})_2O_{3.12}$ (stabilized δ-$Bi_2O_3$) [46], $(ZrO_2)_{0.92}(Y_2O_3)_{0.08}$ (8YSZ) [47], $La_{0.8}Sr_{0.2}Ga_{0.83}Mg_{0.17}O_{2.815}$ (LSGM) [48], $Ce_{0.9}Gd_{0.1}O_{1.95}$, [49] $Ba_3MoNbO_{8.5}$, [52] $La_2Mo_2O_9$, [53] $CsBi_2Ti_2NbO_{10}$, [54] $La_{10}Si_6O_{26}$, [55] and $Ba_7Nb_{3.8}Mo_{1.2}O_{20.1}$ (this work). **b** $\sigma_{bulk}$ of $Pb_{0.8}La_{0.2}WO_{4.1}$, [56] $Bi_{0.9}Sr_{0.1}VO_{3.95}$, [57] $La_{0.9}Na_{0.1}PO_{3.9}$, [58] $La_{0.7}Sr_{2.3}GaO_{4.85}$, [59] $Sr_3V_2O_8$, [51] and $Ba_7Nb_{3.8}Mo_{1.2}O_{20.1}$ (this work). **c** $\sigma_{O^{2-}}$ of $Ce_{0.8}La_{0.2}NbO_{4+\delta}$, [60] and $Ba_7Nb_{3.8}Mo_{1.2}O_{20.1}\cdot y\ H_2O$ (this work), which are calculated by $\sigma_{O^{2-}} = \sigma_{tot} \times t_{O^{2-}}$. Copyright© 2023 Authors

and 4926 times higher compared to $Ba_3YGa_2O_{7.5}$; [65] Fig. 3.15b). $Ba_7Nb_4MoO_{20}\cdot y\ H_2O$ is a candidate material for a dual-ion solid electrolyte in a ceramic fuel cell [13]. Therefore, $Ba_7Nb_{3.8}Mo_{1.2}O_{20.1}\cdot y\ H_2O$ has a higher potential as a dual-ion solid electrolyte in a ceramic fuel cell because of its exceedingly high conductivity at low temperatures, high chemical and electrical stability, phase stability, and absence of expensive rare earth elements and toxic Pb.

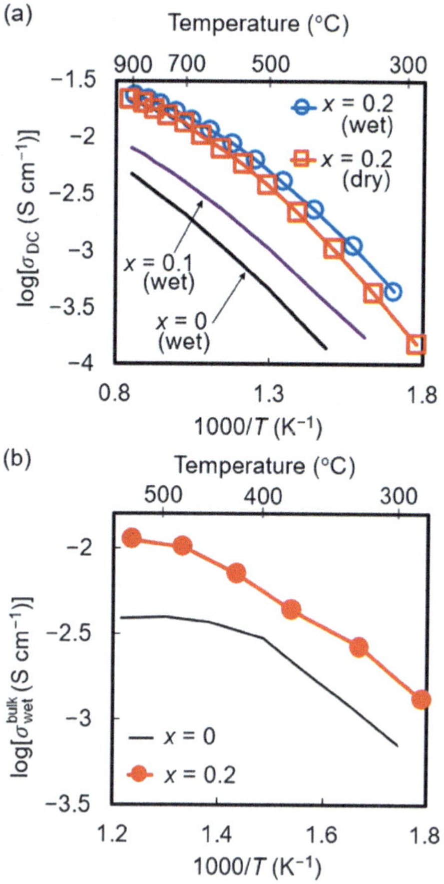


**Fig. 3.14** Proton conduction of $Ba_7Nb_{4-x}Mo_{1+x}O_{20+x/2}\cdot yH_2O$ ($x = 0$, 0.1 and 0.2). **a** Arrhenius plots of DC electrical conductivity $\sigma_{DC}$ in wet air (vapor partial pressure, $P(H_2O) = 0.021$ atm) of $Ba_7Nb_{4-x}Mo_{1+x}O_{20+x/2}\cdot yH_2O$ ($x = 0$ (black line) [18], $x = 0.1$ (purple line) [15], $x = 0.2$; blue open circles and line). $\sigma_{DC}$ for $x = 0.2$ in dry air is also plotted (red line and open squares, this work). **b** Arrhenius plots of bulk conductivities of $Ba_7Nb_{4-x}Mo_{1+x}O_{20+x/2}\cdot yH_2O$ ($x = 0$ (black line) [13], 0.2 (red closed circles and line) (this work)) in wet air ($H_2O$ partial pressure, $P(H_2O) = 0.021$ atm). Copyright© 2023 Authors

### 3.3.4 *Mechanisms of the Stabilization of Excess Oxygen Atoms and High Oxide-Ion Conduction in $Ba_7Nb_{3.8}Mo_{1.2}O_{20.1}$*

#### 3.3.4.1 Crystal Structure Analyses of $Ba_7Nb_{3.8}Mo_{1.2}O_{20.1}$ and $Ba_7Nb_4MoO_{20}$ at 800 °C

To clarify the structural origin of the high oxide-ion conduction in $Ba_7Nb_{3.8}Mo_{1.2}O_{20.1}$, Rietveld refinements and maximum-entropy method (MEM) analyses were performed for neutron-diffraction data of $Ba_7Nb_{3.8}Mo_{1.2}O_{20.1}$ and $Ba_7Nb_4MoO_{20}$ measured in situ at 800 °C. The neutron-diffraction data were analyzed by the $P\bar{3}m1$ 7H polytype (Figs. 3.16 and 3.17; Tables 3.2 and 3.3). Since Mo and Nb atoms have quite similar neutron scattering lengths, they were assumed to be disordered at each Nb/Mo$i$ cation site ($i = 1, 2,$ and 3) in the Rietveld and MEM

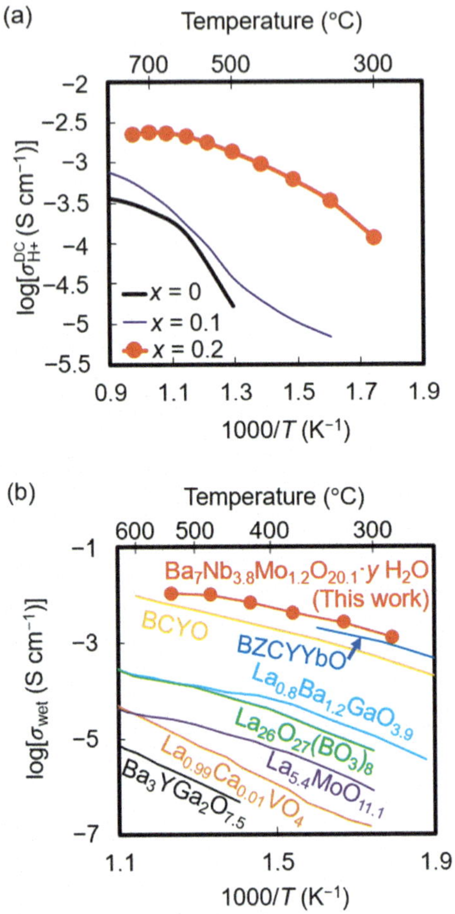


**Fig. 3.15** Proton conduction of $Ba_7Nb_{4-x}Mo_{1+x}O_{20+x/2}\cdot y$ $H_2O$ ($x$ = 0, 0.1 and 0.2). It should be noted that the proton conductivity is estimative, because the oxide-ion conductivity in wet air might be higher than that in dry air. **a** Temperature dependence of total DC proton conductivities of $Ba_7Nb_{4-x}Mo_{1+x}O_{20+x/2}\cdot y$ $H_2O$ ($x$ = 0 (black line) [18], $x$ = 0 (purple line) [15] and 0.2 (red closed circles and line) (this work)). The proton conductivities were estimated from the difference between the conductivities in wet and in dry air. **b** Bulk conductivity of $Ba_7Nb_{3.8}Mo_{1.2}O_{20.1}\cdot y$ $H_2O$ in wet air $\sigma_{wet}^{bulk}$ ($H_2O$ partial pressure, $P(H_2O) = 0.021$ atm; red circles) compared with $La_{0.8}Ba_{1.2}GaO_{3.9}$, [61] $La_{26}O_{27}(BO_3)_8$, [62] $La_{5.4}MoO_{11.1}$, [63] $La_{0.99}Ca_{0.01}VO_4$, [64] $Ba_3YGa_2O_{7.5}$, [65] $BaCe_{0.9}Y_{0.1}O_{3-\delta}$ (BCYO) [66] and $BaZr_{0.1}Ce_{0.7}Y_{0.1}Yb_{0.1}O_{3-\delta}$ [67]. Copyright© 2023 Authors

analyses. The oxygen contents $w$ in $Ba_7Nb_{3.8}Mo_{1.2}O_w$ and $Ba_7Nb_4MoO_w$ were fixed to be $w$ = 20.1 and 20.0, respectively, in the Rietveld analyses of the neutron diffraction data taken at 800 °C, because thermogravimetric (TG) measurements revealed negligibly small weight changes in dry atmospheres at high temperatures of 700–900 °C ($Ba_7Nb_{3.8}Mo_{1.2}O_w$: Fig. 3.18b for $Ba_7Nb_{3.8}Mo_{1.2}O_w$ and Refs. [25, 80] for

$Ba_7Nb_4MoO_w$). Refined occupancy factors, low weighted profile reliability factors $R_{wp}$ and MEM neutron scattering length density (NSLD) distributions indicated the presence of interstitial O5 atoms in the c′ layers of both $Ba_7Nb_{3.8}Mo_{1.2}O_{20.1}$ and $Ba_7Nb_4MoO_{20}$ at 800 °C (Figs. 3.19a–d, 3.20; Tables 3.2, 3.3 and 3.4). The refined occupancy factor of interstitial O5 atom 0.064(11) in $Ba_7Nb_{3.8}Mo_{1.2}O_{20.1}$ is significantly higher than 0.030(3) in $Ba_7Nb_4MoO_{20}$ (Tables 3.2, 3.3) Thus, the excess oxygen atoms $O_{0.1}$ in $Ba_7Nb_{3.8}Mo_{1.2}O_{20.1}$ are incorporated as the interstitial oxygen atoms. The calculated intensities agreed well with the observed intensities giving reasonably small reliability factors. The bond-valence sum (BVS) of each site with the occupancy factor near unity agrees with the average oxidation number (Tables 3.2 and 3.3), validating the refined crystal structures. Figure 3.21shows the refined crystal structure of $Ba_7Nb_{3.8}Mo_{1.2}O_{20.1}$ at 800 °C. The crystal structure of $Ba_7Nb_{3.8}Mo_{1.2}O_{20.1}$ ($x = 0.2$) and $Ba_7Nb_4MoO_{20}$ ($x = 0$) consists of stacking of hexagonal and cubic close-packed $BaO_3$ layers (h and c, respectively), oxygen-deficient cubic close-packed [$Ba1(O1)_{2.1-z}(O5)_z$ for $x = 0.2$ and $Ba1(O1)_{2-z}(O5)_z$ for $x = 0$] (c′) layers and Nb/Mo cations in a c′hhcchh sequence. Here $z = 3g(\text{O}; \text{O5})$, and $g(X; s)$ is the occupancy factor of atom $X$ at the $s$ site. The occupancy factors of all the cation sites and the oxygen O2, O3 and O4 sites are unity forming the h and c close-packed layers. In contrast, the lattice oxygen O1 and interstitial oxygen O5 sites are partially occupied, forming the intrinsically oxygen deficient c′ layer (Fig. 3.21a, b). As discussed below, the oxide ions two-dimensionally migrate in the c′ layer, while the protons migrate in the h and c′ layers.

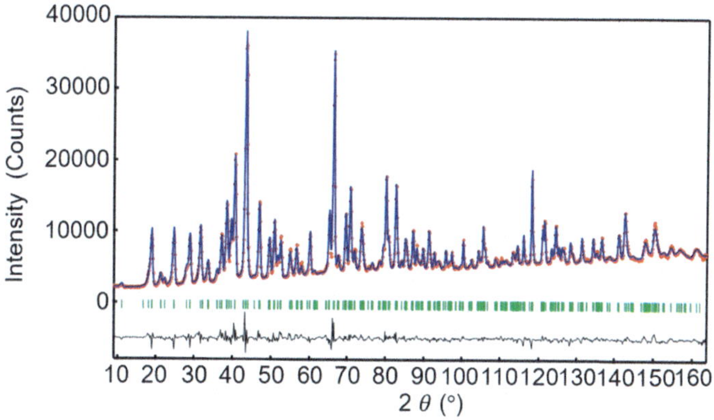


**Fig. 3.16** Rietveld pattern of neutron-diffraction data of $Ba_7Nb_{3.8}Mo_{1.2}O_{20.1}$ taken in situ in vacuum at 800 °C with Echidna diffractometer. Red crosses, dark blue lines, green bars, and black lines represent observed intensities, calculated intensities, peak positions, and difference plots, respectively. Copyright© 2023 Authors

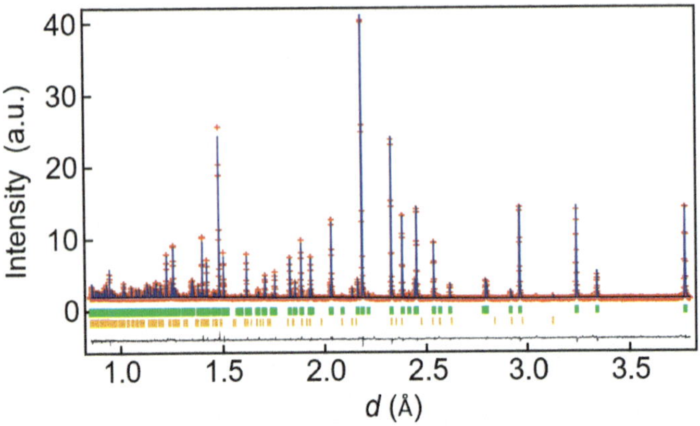


**Fig. 3.17** Rietveld patterns of neutron-diffraction data of $Ba_7Nb_4MoO_{20}$ taken with the backscattering bank ($d = 0.85$–3.8 Å) of the SuperHRPD diffractometer in vacuum at 800 °C. Red crosses, dark blue lines, and black lines represent observed intensities, calculated intensities, and difference plots, respectively. Green tick marks and orange tick marks stand for calculated Bragg peak positions of $Ba_7Nb_4MoO_{20}$ and $Ba_5Nb_4O_{15}$, respectively. Copyright© 2023 Authors

#### 3.3.4.2 Presence of $(Nb/Mo)_2O_9$ Dimers and $(Nb/Mo)O_5$ Monomer

To examine the local dynamic structures, we performed ab initio molecular dynamics (AIMD) simulations at 800 and 1200 °C using 3 × 3 × 1 supercells of $(Ba_7Nb_4MoO_{20})_9$ containing 288 atoms and $Ba_{63}Nb_{34}Mo_{11}O_{181}$ [~ $(Ba_7Nb_{3.8}Mo_{1.2}O_{20.1})_9$] containing 289 atoms. Snapshots and histograms of the coordination number of Nb/Mo*i* cations ($i$ = 1, 2, and 3) from the AIMD simulations indicate that the Nb/Mo1 and Nb/Mo3 atoms mostly take octahedral coordination (Fig. 3.22a, c), which is consistent with the average crystal structure from the Rietveld analysis. In contrast to the average structure, the Nb/Mo2 cation near the ion-conducting c′ layer takes little octahedral coordination [$(Nb/Mo2)O_6$ octahedra] (Fig. 3.22b). It should be noted that the excess oxygen atoms are accommodated in the oxygen-deficient c′ layer, forming $M_2O_9$ dimers and $MO_5$ monomers ($M_2O_9$ and $MO_5$ in Fig. 3.23) where $M$ denotes the Nb/Mo2 atom in the AIMD simulations. In the $M_2O_9$ dimer, two $MO_5$ monomers are connected by a corner-sharing oxygen $O_{sh}$ atom, producing considerable local relaxation of the $MO_5$ monomers in the dimer (see the $M_2O_9$ dimers (squares) in **B** and **C** of Fig. 3.24b. The $MO_5$ monomer is an $O_{ap}'$–$O_{ap}'$ edge-capped trigonal plane (three oxygen atoms) where $O_{ap'}$ is an apical oxygen atom coordinated with an Nb/Mo2 cation in $MO_5$ monomer (Fig. 3.23c). It should be noted that the spatial- and time-averaged probability-density distribution of oxygen atoms in the c′ layer from the local dynamic structures in the AIMD simulations (Fig. 3.19e) is consistent with the MEM NSLD distribution (average structure) from the neutron-diffraction experiments (Fig. 3.19c). The $O_{ap}'$–$O_{ap}'$ repulsion in $MO_5$ results in the delocalized spatial distribution of the $O_{ap'}$ atom (Fig. 3.19e), which is experimentally validated by the large MEM NSLD distribution around the

**Table 3.2** Refined crystallographic parameters and reliability factors in the Rietveld analysis of the neutron-diffraction data of $Ba_7Nb_{3.8}Mo_{1.2}O_{20.1}$ taken with Echidna at 800 °C

| Site *s* Atom label | Atom *X* | Wyckoff position | *g*(*X*; *s*) | *x* | *y* | *z* |
|---|---|---|---|---|---|---|
| Ba1 | Ba | 1*a* | 1 | 0 | 0 | 0 |
| Ba2 | Ba | 2*d* | 1 | 1/3 | 2/3 | 0.8233(5) |
| Ba3 | Ba | 2*d* | 1 | 1 | 2/3 | 0.5739(6) |
| Ba4 | Ba | 2*c* | 1 | 0 | 0 | 0.2859(6) |
| Nb/Mo1 | $Nb_{0.76}Mo_{0.24}$ | 1*b* | 1 | 0 | 0 | 1/2 |
| Nb/Mo2 | $Nb_{0.76}Mo_{0.24}$ | 2*d* | 1 | 1/3 | 2/3 | 0.0927(4) |
| Nb/Mo3 | $Nb_{0.76}Mo_{0.24}$ | 2*d* | 1 | 1/3 | 2/3 | 0.3497(4) |
| O1 | O | 6*i* | 0.318(6) | 0.346(10) | 0.69(2) | − 0.0142(7) |
| O2 | O | 6*i* | 1 | 0.1673(9) | − 0.1673(9) | 0.1307(4) |
| O3 | O | 6*i* | 1 | 0.1660(8) | − 0.1660(8) | 0.4334(3) |
| O4 | O | 6*i* | 1 | 0.4912(6) | − 0.4912(6) | 0.2954(3) |
| O5 | O | 3*e* | 0.064(11) | 1/2 | 0 | 0 |

| Site *s* Atom label | $U_{iso}$ or $U_{eq}$ (Å$^2$) | $U_{11}$ (Å$^2$) | $U_{22}$ (Å$^2$) | $U_{33}$ (Å$^2$) | $U_{12}$ (Å$^2$) | $U_{13}$ (Å$^2$) | $U_{23}$ (Å$^2$) | BVS (vu) |
|---|---|---|---|---|---|---|---|---|
| Ba1 | 0.042(7) | 0.051(7) | 0.051(7) | 0.024(8) | 0.029(4) | 0 | 0 | 1.86 |
| Ba2 | 0.0273(11) | | | | | | | 1.98 |
| Ba3 | 0.0273(11) | | | | | | | 2.17 |
| Ba4 | 0.0273(11) | | | | | | | 1.75 |
| Nb/Mo1 | 0.0247(6) | | | | | | | 4.12 |
| Nb/Mo2 | 0.0247(6) | | | | | | | 4.48 |
| Nb/Mo3 | 0.0247(6) | | | | | | | 4.41 |
| O1 | 0.085(11) | | | | | | | 1.65 |
| O2 | 0.059(4) | 0.063(4) | 0.063(4) | 0.078(4) | 0.052(4) | 0 | 0 | 1.70 |
| O3 | 0.035(2) | 0.032(2) | 0.032(2) | 0.038(2) | 0.014(3) | 0 | 0 | 1.81 |

(continued)

**Table 3.2** (continued)

| Site $s$ Atom label | $U_{iso}$ or $U_{eq}$ ($Å^2$) | $U_{11}$ ($Å^2$) | $U_{22}$ ($Å^2$) | $U_{33}$ ($Å^2$) | $U_{12}$ ($Å^2$) | $U_{13}$ ($Å^2$) | $U_{23}$ ($Å^2$) | BVS (vu) |
|---|---|---|---|---|---|---|---|---|
| O4 | 0.041(3) | 0.045(3) | 0.045(3) | 0.055(4) | 0.039(3) | 0 | 0 | 1.86 |
| O5 | 0.085(11) | | | | | | | 1.21 |

Bravais lattice: hexagonal: Crystal system: trigonal. Space group: $P\overline{3}m1$. Lattice parameters: $a = b = 5.92920(19)$ Å, $c = 16.6994(5)$ Å. Number of formula per unit cell: $Z = 1$
$g(X; s)$: Occupancy factor of $X$ atom at the $s$ site. Linear constraints: $g$(Ba; Ba1) = $g$(Ba; Ba2) = $g$(Ba; Ba3) = $g$(Ba; Ba4) = $g$($Nb_{0.76}Mo_{0.24}$; Nb/Mo1) = $g$($Nb_{0.76}Mo_{0.24}$; Nb/Mo2) = $g$($Nb_{0.76}Mo_{0.24}$; Nb/Mo3) = $g$(O; O2) = $g$(O; O3) = $g$(O; O4) = 1; 6 $g$(O; O1) + 3 $g$(O; O5) = 2.1
$U_{iso}(Xn)$: Isotropic atomic displacement parameter of $X$ atom at the $Xn$ site. Linear constraints: $U_{iso}$(Ba2) = $U_{iso}$(Ba3) = $U_{iso}$(Ba4), $U_{iso}$(Nb/Mo1) = $U_{iso}$(Nb/Mo2) = $U_{iso}$(Nb/Mo3), $U_{iso}$(O1) = $U_{iso}$(O5).
Reliability factors: $R_{wp} = 4.18\%$, $R_p = 3.11\%$, $R_B = 1.08\%$, $R_F = 0.54\%$
BVS: bond valence sum. BVSs were calculated using the bond-valence parameters reported by Gagné and Hawthorne [68]. Low BVS value at the interstitial O5 site 1.21 indicates that the O5 oxide ion is underbonded. Such underbonding at the interstitial site has been observed also in other compositions such as $Ba_7Nb_{3.9}Mo_{1.1}O_{20.05}$ [15]
In preliminary analyses, the refined occupancy factors of Ba atoms were 1.00 or higher than 1.00, thus, we fixed them to unity in the final refinement

apical O1 atom (Fig. 3.19c) and high O1 atomic displacement parameter (Table 3.2). Similar (Nb/Mo)$O_5$ polyhedra were suggested by total neutron scattering experiments on $Ba_3MoNbO_{8.5}$ [36] and by AIMD simulation of $Ba_7Nb_4MoO_{20.5}H$ with excess oxygen atom $O_{0.5}$ [69].

#### 3.3.4.3 Oxide-Ion Migration Between the Lattice O1 and Interstitial O5 Sites

Next, we discuss the oxide-ion migration in $Ba_7Nb_{3.8}Mo_{1.2}O_{20.1}$ and $Ba_7Nb_4MoO_{20}$. The mean-square displacements (MSDs) of Ba, Nb, and Mo in both $Ba_7Nb_{3.8}Mo_{1.2}O_{20.1}$ and $Ba_7Nb_4MoO_{20}$ in the AIMD simulations indicate no cation diffusion (Fig. 3.25a), which is consistent with both the isolated MEM NSLD distributions of cations (Fig. 3.19a–c) and high oxide-ion transport number (Fig. 3.7a). The connected NSLD distribution between the O1 and O5 sites in $Ba_7Nb_{3.8}Mo_{1.2}O_{20.1}$ (Fig. 3.19c) exhibits good agreement with the connected average probability distribution of oxide ions between the O1 and O5 sites from AIMD simulation for $Ba_7Nb_{3.8}Mo_{1.2}O_{20.1}$ (Fig. 3.19e), which indicates the oxide-ion migration pathway along the $[1\overline{1}0]$, [120], and $[\overline{2}\overline{1}0]$ directions at 800 °C (Fig. 3.19d). Similarly, the connected NSLD distributions are observed between the O1 and O5 sites also in $Ba_7Nb_4MoO_{20}$, (Fig. 3.19a). The minimum NSLD between the O1 and O5 sites of $Ba_7Nb_{3.8}Mo_{1.2}O_{20.1}$ is higher than that of $Ba_7Nb_4MoO_{20}$ (Fig. 3.19b, c), which is consistent with the higher bulk conductivity of $Ba_7Nb_{3.8}Mo_{1.2}O_{20.1}$ (Fig. 3.7d). The average probability density distributions and trajectories of oxide

**Table 3.3** Refined crystallographic parameters and reliability factors in the Rietveld analysis of the neutron-diffraction data of $Ba_7Nb_4MoO_{20}$ taken with SuperHRPD at 800 °C

| Site *s* Atom label | Atom *X* | Wyckoff position | *g*(*X*; *s*) | *x* | *y* | *z* |
|---|---|---|---|---|---|---|
| Ba1 | Ba | 1*a* | 1 | 0 | 0 | 0 |
| Ba2 | Ba | 2*d* | 1 | 1/3 | 2/3 | 0.82341(12) |
| Ba3 | Ba | 2*d* | 1 | 1 | 2/3 | 0.57479(15) |
| Ba4 | Ba | 2*c* | 1 | 0 | 0 | 0.28143(15) |
| Nb/Mo1 | $Nb_{0.8}Mo_{0.2}$ | 1*b* | 1 | 0 | 0 | 0.5 |
| Nb/Mo2 | $Nb_{0.8}Mo_{0.2}$ | 2*d* | 1 | 1/3 | 2/3 | 0.09382(9) |
| Nb/Mo3 | $Nb_{0.8}Mo_{0.2}$ | 2*d* | 1 | 1/3 | 2/3 | 0.34973(10) |
| O1 | O | 6*i* | 0.3183(14) | 0.3658(5) | 0.7316(10) | – 0.01228(13) |
| O2 | O | 6*i* | 1 | 0.1669(2) | – 0.1669(2) | 0.12954(9) |
| O3 | O | 6*i* | 1 | 0.1643(2) | – 0.1643(2) | 0.43205(8) |
| O4 | O | 6*i* | 1 | 0.49228(16) | – 0.49228(16) | 0.29585(7) |
| O5 | O | 3*e* | 0.030(3) | ½ | 0 | 0 |

| Site *s* Atom label | $U_{iso}$ or $U_{eq}$ (Å$^2$) | $U_{11}$ (Å$^2$) | $U_{22}$ (Å$^2$) | $U_{33}$ (Å$^2$) | $U_{12}$ (Å$^2$) | $U_{13}$ (Å$^2$) | $U_{23}$ (Å$^2$) | BVS (vu) |
|---|---|---|---|---|---|---|---|---|
| Ba1 | 0.0353(12) | 0.0378(18) | 0.0378(18) | 0.030(3) | 0.0189(9) | 0 | 0 | 1.91 |
| Ba2 | 0.0208(3) | | | | | | | 1.95 |
| Ba3 | 0.0208(3) | | | | | | | 2.15 |
| Ba4 | 0.0208(3) | | | | | | | 1.74 |
| Nb/Mo1 | 0.0133(2) | | | | | | | 4.18 |
| Nb/Mo2 | 0.0133(2) | | | | | | | 4.50 |
| Nb/Mo3 | 0.0133(2) | | | | | | | 4.40 |
| O1 | 0.0332(16) | | | | | | | 1.62 |
| O2 | 0.0494(6) | 0.0534(8) | 0.0534(8) | 0.0413(10) | 0.0418(10) | 0 | 0 | 1.74 |
| O3 | 0.0227(4) | 0.0240(6) | 0.0240(6) | 0.0201(7) | 0.0133(7) | 0 | 0 | 1.79 |

(continued)

**Table 3.3** (continued)

| Site *s* Atom label | $U_{iso}$ or $U_{eq}$ (Å$^2$) | $U_{11}$ (Å$^2$) | $U_{22}$ (Å$^2$) | $U_{33}$ (Å$^2$) | $U_{12}$ (Å$^2$) | $U_{13}$ (Å$^2$) | $U_{23}$ (Å$^2$) | BVS (vu) |
|---|---|---|---|---|---|---|---|---|
| O4 | 0.0296(5) | 0.0298(7) | 0.0298(7) | 0.0294(8) | 0.0238(8) | 0 | 0 | 1.86 |
| O5 | 0.0332(16) | | | | | | | 1.20 |

Bravais lattice: hexagonal: Crystal system: trigonal. Space group: $P\overline{3}m1$. Lattice parameters: $a = b =$ 5.931455(3) Å, $c =$ 16.70198(2) Å. Number of formula per unit cell: $Z = 1$
$g(X; s)$: Occupancy factor of $X$ atom at the $s$ site. $g$(Ba; Ba1) = $g$(Ba; Ba2) = $g$(Ba; Ba3) = $g$(Ba; Ba4) = $g$($Nb_{0.8}Mo_{0.2}$; Nb/Mo1) = $g$($Nb_{0.76}Mo_{0.24}$; Nb/Mo2) = $g$($Nb_{0.8}Mo_{0.2}$; Nb/Mo3) = $g$(O; O2) = $g$(O; O3) = $g$(O; O4) = $g$($Nb_{0.8}Mo_{0.2}$; Nb/Mo2) = 1; 6 $g$(O; O1) + 3 $g$(O; O5) = 2
$U_{iso}(Xn)$: Isotropic atomic displacement parameter $Xn$ atom
Linear constraints: $U_{iso}$(Ba2) = $U_{iso}$(Ba3) = $U_{iso}$(Ba4), $U_{iso}$(Nb/Mo1) = $U_{iso}$(Nb/Mo2) = $U_{iso}$(Nb/Mo3), $U_{iso}$(O1) = $U_{iso}$(O5)
Reliability factors: $R_{wp} = 2.82\%$, $R_p = 2.13\%$, $R_B = 4.46\%$, $R_F = 7.59\%$
Mass ratio; $Ba_7Nb_4MoO_{20}$: $Ba_5Nb_4O_{15}$ = 0.9705(10): 0.0295(10)
BVS: bond valence sum. BVSs were calculated using the bond-valence parameters reported by Gagné and Hawthorne [68]. Low BVS value at the interstitial O5 site 1.20 indicates that the O5 oxide ion is underbonded. Such underbonding at the interstitial site has been observed also in other compositions such as $Ba_7Nb_{3.9}Mo_{1.1}O_{20.05}$ [15]

ions from AIMD also support this difference between the two compositions. The average probability density distribution between O1 and O5 sites is not connected in $Ba_7Nb_4MoO_{20}$ but in $Ba_7Nb_{3.8}Mo_{1.2}O_{20.1}$ they are connected (Fig. 3.26). The oxide-ion trajectories indicate (i) only short-range oxide-ion migration and vibration around each oxygen site in $Ba_7Nb_4MoO_{20}$ and (ii) long-range oxide-ion O1–O5 migration in $Ba_7Nb_{3.8}Mo_{1.2}O_{20.1}$ (Fig. 3.27). The two-dimensional oxide-ion O1–O5 migration in the c′ layer is supported by (i) bond-valence-based energy landscape (Fig. 3.28(b)), (ii) average probability density distributions, trajectories of oxide ions and MSDs of oxide ions along *a*, *b* and *c* axes from AIMD simulations (Figs. 3.25b, 3.26). These results indicate that the high oxide-ion conduction in $Ba_7Nb_{3.8}Mo_{1.2}O_{20.1}$ is ascribed to the oxide-ion migration between the apical O1 and interstitial O5 sites in the oxygen-deficient c′ layer. Similar two-dimensional diffusion pathways, where oxide ions migrate through apical and interstitial oxygens, have been found in other hexagonal perovskite-related oxides such as $Ba_3MoNbO_{8.5}$ and $Ba_7Nb_{3.9}Mo_{1.1}O_{20.05}$, suggesting that this is a common feature in the hexagonal perovskite-related family [4, 15, 17, 18, 25, 70, 71]. Indeed, the bond-valence-based energy landscape of $Ba_3MoNbO_{8.5}$ is similar with those of $Ba_7Nb_{3.8}Mo_{1.2}O_{20.1}$ and $Ba_7Nb_4MoO_{20}$ (Fig. 3.28).

#### 3.3.4.4 $M_2O_9$ Dimer-Mediated Oxide-Ion Diffusion Mechanism

To examine the local dynamics of the oxide-ion migration process, we investigated the snapshots from AIMD simulations. Figure 3.24shows a local oxide-ion migration process, which takes only 845 fs. The oxide ions migrate by the breaking and

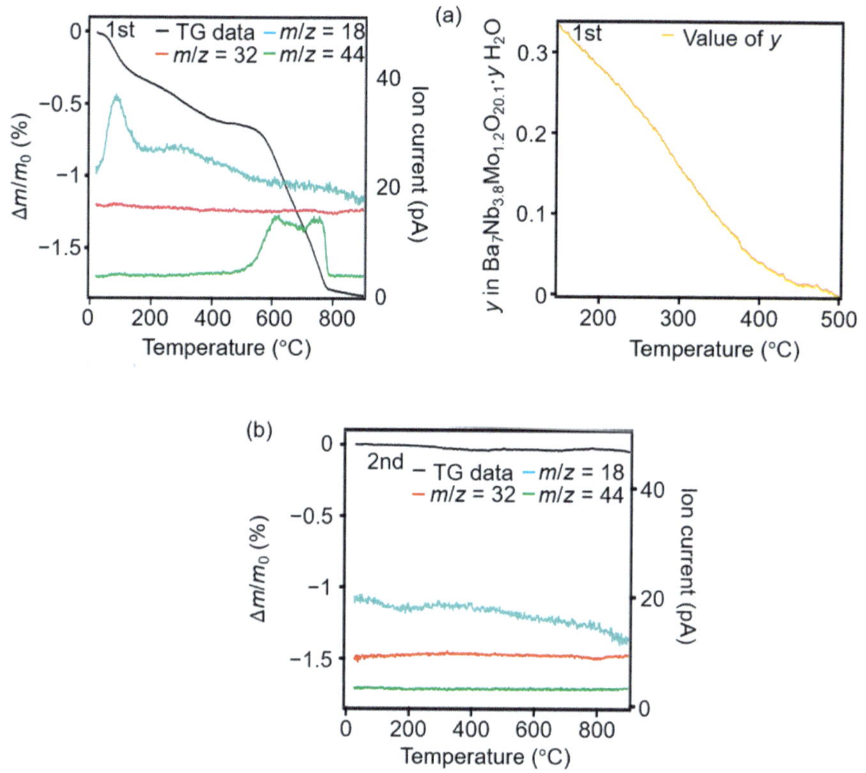


**Fig. 3.18** Thermogravimetry-mass spectroscopy (TG-MS) data of $Ba_7Nb_{3.8}Mo_{1.2}O_{20.1}{\cdot}y\ H_2O$ on **a** 1st and **b** 2nd heating under He gas flow. $\Delta m/m_0 = 100(m - m_0)/m_0$ (%) where the $m$ denotes the sample weight at a temperature and $m_0$ represents the sample weight at **a** 22 °C and **b** 27 °C. The right figure of panel (a) shows the temperature dependence of water content $y$ in $Ba_7Nb_{3.8}Mo_{1.2}O_{20.1}{\cdot}y\ H_2O$, which indicates $y = 0.33$ at 150 °C. Below 150 °C, the weight loss on heating can be mainly ascribed to the desorption of absorbed water at the sample surface, thus, the water content in the lattice would approximately be $y = 0.33$ also at RT in the as-prepared $Ba_7Nb_{3.8}Mo_{1.2}O_{20.1}{\cdot}y\ H_2O$ sample. Black line denotes the TG data, and the light blue, red, and green lines stand for the ion current data for $m/z = 18$, $m/z = 32$, and $m/z = 44$, respectively. Copyright© 2023 Authors

reforming of $M_2O_9$ dimers, $MO_5$ monomers and $MO_4$ tetrahedra. It should be noted that the breaking and reforming of $M_2O_9$ dimers are essential for oxide-ion migration. For example, $MO_5$ [= $(M\text{a})O_5$] and $MO_4$ [= $(M\text{b})O_4$] in panel **A** in Fig. 3.24a–c break between 0 and 166 fs, forming a dimer $M_2O_9$ [= $(M\text{a})(M\text{b})O_9$] in panel **B** in Fig. 3.24a–c. The $(M\text{a})(M\text{b})O_9$ dimer and $(M\text{c})O_4$ tetrahedron break between 166 and 660 fs forming $(M\text{a})O_4$ tetrahedron and $(M\text{b})(M\text{c})O_9$ dimer in panel **C** in Fig. 3.24a–c. Between 660 and 845 fs, the $(M\text{a})(M\text{b})O_9$ dimer breaks forming $(M\text{b})O_4$ tetrahedron and $(M\text{c})O_5$ monomer in panel **D** in Fig. 3.24a–c. During this process, the red Ob atom migrates from the apical $O_{ap}$' site of the $M\text{a}O_5$ monomer to the nearest-neighbor $O_{ap}$ site of the $M\text{b}O_4$ tetrahedron through the corner-sharing

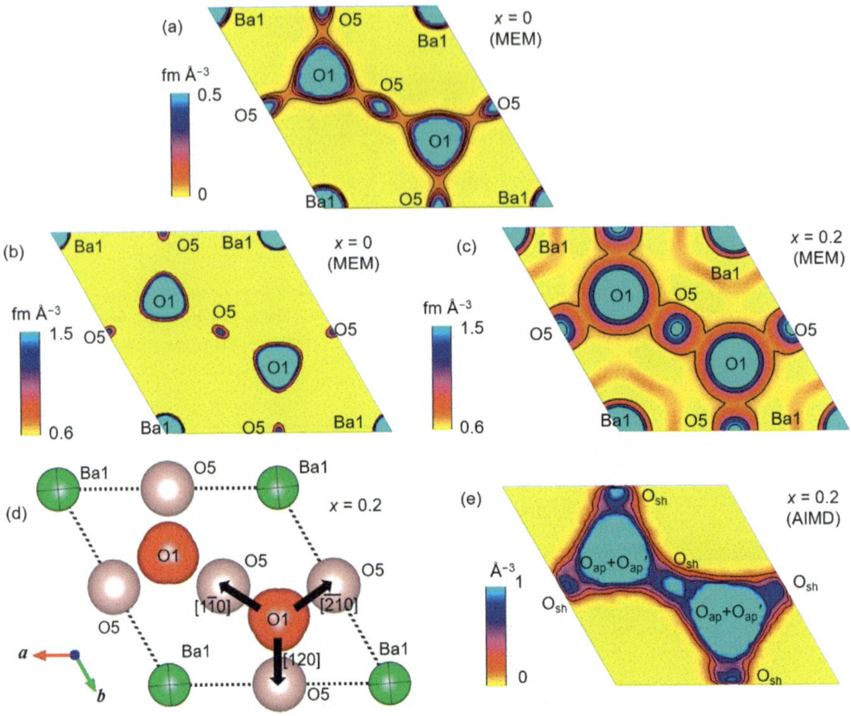


**Fig. 3.19** Presence of interstitial oxygen atom O5, disorder, diffusion pathway, and interstitialcy diffusion of oxide ions in $Ba_7Nb_{3.8}Mo_{1.2}O_{20.1}$ and $Ba_7Nb_4MoO_{20}$ at 800 °C. **a, b, c** MEM NSLDs on the *ab* plane at $z = 0$ of $Ba_7Nb_{4-x}Mo_{1+x}O_{20+x/2}$ ($x = 0, 0.2$). The contour lines are from (a) 0 to 0.5 fm Å$^{-3}$ by the step of 0.1 fm Å$^{-3}$ and (b, c) 0.6 to 1.5 fm Å$^{-3}$ by the step of 0.4 fm Å$^{-3}$. Compositions are (a, b), $x = 0$ and (c), $x = 0.2$. **d** Refined crystal structure on the *ab* plane at $z = 0$ of $Ba_7Nb_{3.8}Mo_{1.2}O_{20.1}$ using neutron-diffraction data. Thermal ellipsoids are drawn at the 90% probability level. Arrows denote the directions of oxide-ion O1-to-O5 migration. **e** Corresponding spatial probability-density distribution of oxide ions on the *ab* plane in $(Ba_7Nb_{3.8}Mo_{1.2}O_{20.1})_9$ calculated by AIMD simulations, where the probability density distribution in the $3 \times 3 \times 1$ supercell is averaged to $1 \times 1 \times 1$. The contour lines are from 0 to 1 with step 0.2. Copyright© 2023 Authors

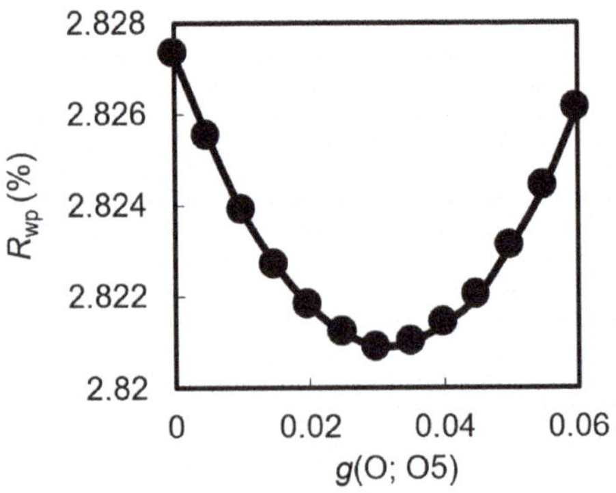


**Fig. 3.20** Variation of the reliability factor $R_{wp}$ with the occupancy factor $g$(O; O5) in the Rietveld analyses of neutron-diffraction data of $Ba_7Nb_4MoO_{20}$ taken with SuperHRPD at 800 °C, showing the validity of refined $g$(O; O5) value, 0.030. Copyright© 2023 Authors

**Table 3.4** Comparison of the reliability factors between the structural models with and without the interstitial O5 atoms at 800 °C

| Composition | Model with O5 atoms | | | Model without O5 atoms | |
|---|---|---|---|---|---|
| | $g$(O; O5) | $R_{wp}$ % | $R_F$ % | $R_{wp}$ % | $R_F$ % |
| $Ba_7Nb_{3.8}Mo_{1.2}O_{20.1}$ | 0.064(11) | 4.18 | 0.54 | 4.20 | 0.55 |
| $Ba_7Nb_4MoO_{20}$ | 0.030(3) | 2.82 | 7.58 | 2.83 | 7.60 |

The values of $R_{wp}$ and $R_F$ for the model with O5 atoms where the $g$(O; O5) was refined are lower than those for the model without O5 atoms where the $g$(O; O5) was fixed to be 0

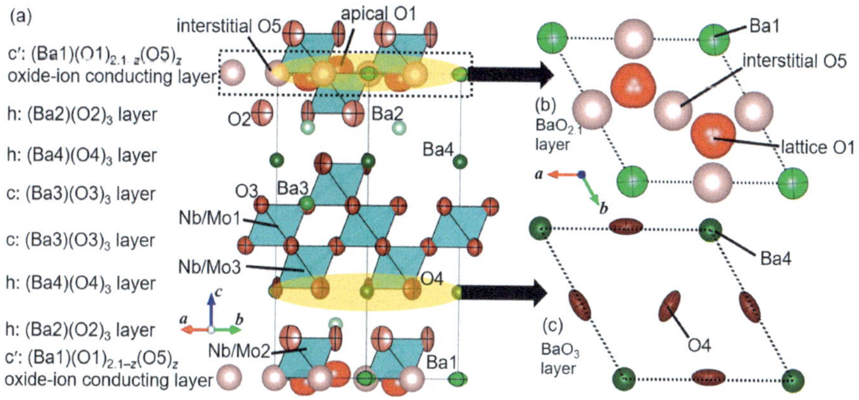


**Fig. 3.21** **a** Refined crystal structure of $Ba_7Nb_{3.8}Mo_{1.2}O_{20.1}$ at 800 °C, which was obtained by the Rietveld analysis. $z$ in $(Ba1)(O1)_{2.1-z}(O5)_z$ stands for the number of O5 oxygen atoms. O5 is the highly defective interstitial oxygen site. Thermal ellipsoids are drawn at the 90% probability level. **b** Intrinsically oxygen deficient $BaO_{2.1}$ [= $(Ba1)(O1)_{2.1-z}(O5)_z$] and **c** fully occupied $BaO_3$ layers in $Ba_7Nb_{3.8}Mo_{1.2}O_{20.1}$. Copyright© 2023 Authors

oxygen $O_{sh}$ site in ($M$a)($M$b)$O_9$ dimer. Similarly, the green Oc atom at the apical $O_{ap}$ site of ($M$b)$O_4$ tetrahedron moves to the apical $O_{ap}'$ site of $M$c$O_5$ monomer through the sharing $O_{sh}$ site in ($M$b)($M$c)$O_9$ dimer. The oxide-ion migration is reminiscent of a concerted push–pull interstitialcy "bucket-relay"-type motions where many people ($M$ cations) line up and hand buckets of water (oxide ions) to their neighbors one after another (Fig. 3.29). It should be noted that the $M_2O_9$ dimer is not a transition state, but stable, because (1) the fraction of $M_2O_9$ dimers (> 90%) is much higher than that of $MO_5$ monomers (< 10%) and (2) the life time of $M_2O_9$ is relatively long compared with $MO_5$. In contrast, a $Si_2O_9$ unit was reported to be likely a transition state in the simulation of oxygen-excess lanthanum silicate apatite [72]. The present "bucket-relay"-type motion is different from the "cog-wheel"- and "paddle-wheel"-type motions [26, 72–75], because the rotational motions between O1 ($O_{ap}$ and $O_{ap}'$) and O2 atoms are quite rare, and the oxide-ion migration is not accompanied by rotational motion in our AIMD simulations of $Ba_7Nb_{3.8}Mo_{1.2}O_{20.1}$ and $Ba_7Nb_4MoO_{20}$ (Figs. 3.24, 3.26, 3.27).

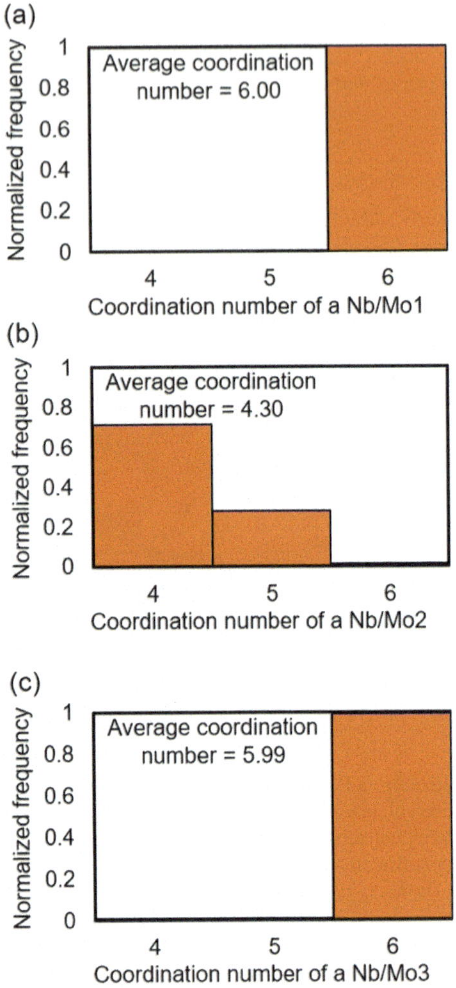


**Fig. 3.22** Histograms of the coordination number of **a** Nb/Mo1, **b** Nb/Mo2, and **c** Nb/Mo3 cations in $Ba_7Nb_{3.8}Mo_{1.2}O_{20.1}$ calculated by ab initio molecular dynamics (AIMD) simulations at 1200 °C. Coordination numbers in the 3 × 3 × 1 supercell was examined from 0 to 74,264 fs. Copyright© 2023 Authors

As discussed above, the excess oxygen defects are accommodated by the formation of $M_2O_9$ dimers and $MO_5$ monomers near the oxygen-deficient c′ layer (Figs. 3.21, 3.22, 3.23 and 3.24). Since the $M2$–O5 distance 2.28 Å in the average structure from neutron diffraction experiments at 800 °C is in agreement with the distance between the $M2$ and corner-sharing oxygen atoms in the local dynamic structure at 800 °C (1.91–2.35 Å), the interstitial oxygen atom O5 corresponds to the corner-sharing oxygen $O_{sh}$ atom of the $M_2O_9$ dimer (Fig. 3.21c, e). Therefore, the presence of interstitial O5 atoms by the Rietveld and MEM analyses also supports the presence of $M_2O_9$ dimer in $Ba_7Nb_{3.8}Mo_{1.2}O_{20.1}$ and $Ba_7Nb_4MoO_{20}$. Since the oxide ion migrates by the breaking and reforming of an $M_2O_9$ dimer, the number of the interstitial oxygen O5 atom would be a measure of carrier concentration. In fact, the bulk conductivity at each temperature increases with an increase of the number of the

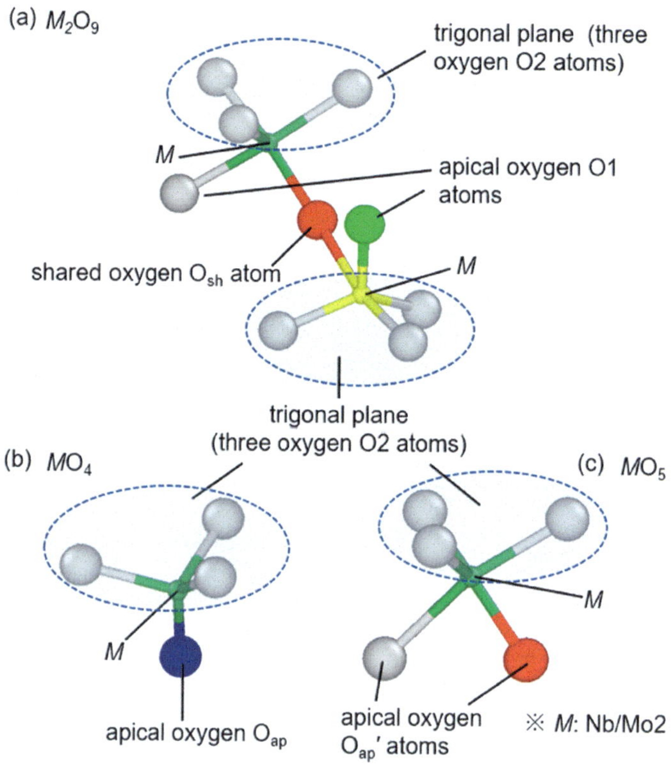


**Fig. 3.23** **a** $M_2O_9$ dimer, **b** $MO_4$ tetrahedron and **c** $MO_5$ monomer near the oxygen-deficient c′ layer in $Ba_7Nb_{3.8}Mo_{1.2}O_{20.1}$ (AIMD calculations at 1200 °C). Three O2 atoms forming a trigonal plane of a polyhedron are relatively rigid, while the apical oxygen $O_{ap}$ atom in $MO_4$, and $O_{ap}'$ atoms in $MO_5$, and corner-sharing oxygen $O_{sh}$ atom in $M_2O_9$ dimer are mobile. These figures show considerable local relaxation associated with an excess oxygen to form an $MO_5$ monomer or an $M_2O_9$ dimer. Copyright© 2023 Authors

O5 atom in a unit cell and amount of excess oxygen atom (Fig. 3.30a, b). This correlation is validated by the minimum NSLD on the O1–O5 pathway. With an increase in the amount of interstitial O5 atoms and excess oxygen atoms, the minimum NSLD on the O1–O5 pathway increases (Fig. 3.30c, d), which leads to higher conductivity (Fig. 3.30e). Similar findings have been reported for $Pr_2NiO_4$-based materials. $(Pr_{0.9}La_{0.1})_2(Ni_{0.74}Cu_{0.21}Ga_{0.05})O_{4+\delta}$ exhibits higher oxygen permeability and oxide-ion conductivity than $Pr_2Ni_{0.75}Cu_{0.25}O_{4+\delta}$, which has been attributed to the higher minimum NSLD on the diffusion pathway and the larger amount of interstitial oxygens created by replacing $Ni^{2+}$ with $Ga^{3+}$ [22, 76].

Several materials were reported to exhibit oxide-ion conduction by the breaking and reforming of dimers (Fig. 3.13b, c) [51, 56–60]. Compared to these material, the present $Ba_7Nb_{3.8}Mo_{1.2}O_{20.1}$ exhibits higher oxide-ion conductivity below 600 °C. Low activation energies (0.25(4)–0.41(4) eV for bulk conductivity and 0.168(10) eV

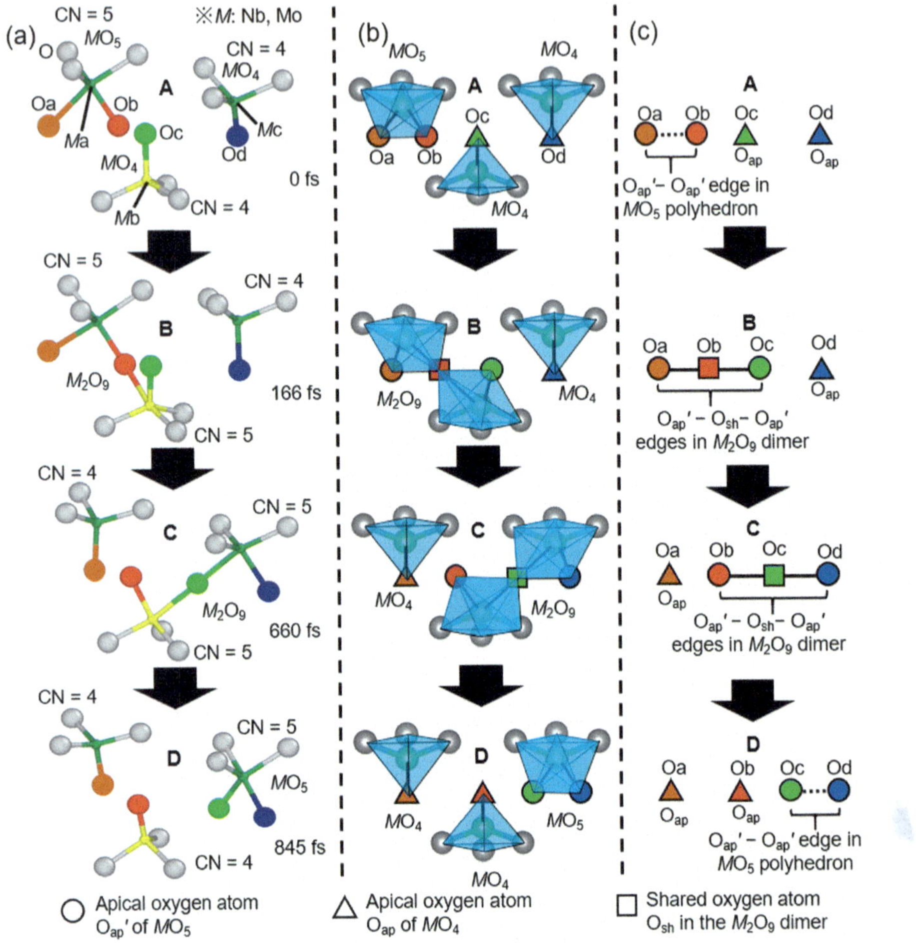


**Fig. 3.24** Snapshots (**a**) and schematic representations (**b**, **c**) of oxygen diffusion process of $Ba_7Nb_{3.8}Mo_{1.2}O_{20.1}$ at 1200 °C. Elapsed times from the start **A** are **B** 166, **C** 660, and **D** 845 fs. The deep green (*M*a, *M*c) and yellow (*M*b) spheres represent Nb or Mo cations at the Nb/Mo2 site. The brown (Oa), red (Ob), light green (Oc), and blue (Od) spheres represent mobile oxide ions in the oxide-ion conducting c′ layer. Gray O2 spheres denote the immobile oxide ions forming the trigonal plane. *M*–O bonds at or less than 2.9 Å are drawn where *M* denotes Nb or Mo cation at the Nb/Mo2 site. In the panels (b) and (c), the circle $O_{ap}'$, triangle $O_{ap}$, square $O_{sh}$, solid line, and dotted line represent an apical oxygen atom of $MO_5$, an apical oxygen atom of $MO_4$, a corner-sharing oxygen atom in $M_2O_9$ dimer, $O_{ap}'$– $O_{sh}$–$O_{ap}'$ edges of an $M_2O_9$ dimer, and $O_{ap}'$–$O_{ap}'$ edge of an $MO_5$ monomer, respectively. The apical $O_{ap}'$ atom is displaced from the regular position due to the repulsion between $O_{ap}'$ and ($O_{ap}'$ or $O_{sh}$) atoms. To emphasize the displacement, $O_{ap}'$ is used for $O_{ap}$ in panel (c). Copyright© 2023 Authors

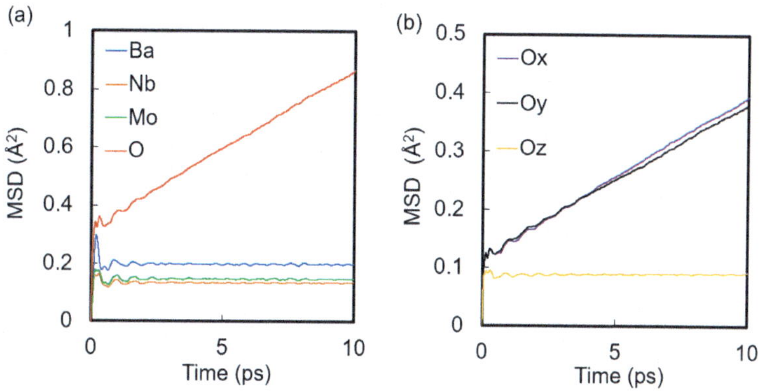


**Fig. 3.25** Mean-square displacement (MSD) of $Ba_7Nb_{3.8}Mo_{1.2}O_{20.1}$ in the AIMD simulations at 1200 °C. **a** MSDs of Ba, Nb, Mo, and O atoms. **b** MSDs of oxygen atoms along the *a* axis (O*x*), *b* axis (O*y*), and *c* axis (O*z*). Copyright© 2023 Authors

for total DC conductivity) of $Ba_7Nb_{3.8}Mo_{1.2}O_{20.1}$ is an origin of the high conductivity at lower temperatures (Fig. 3.13b, c). The short distance between the O1 and O5 atoms 1.7 Å in the refined average crystal structure of $Ba_7Nb_{3.8}Mo_{1.2}O_{20.1}$ would be a reason for the low activation energies, while longer paths are needed for an oxide ion to across the unit cell in other materials in (Fig. 3.13b, c).

#### 3.3.4.5 Origin of Low Activation Energy for Oxide-Ion Conductivity of $Ba_7Nb_{4-x}Mo_{1+x}O_{20+x/2}$

Although the hexagonal perovskite-related oxides $Ba_3MM'O_{8.5}$ ($M$ = Mo, W; $M'$ = V, Nb) have short oxide-ion migration paths, their activation energies for oxide-ion conductivity $E_a$ are much higher than those of $Ba_7Nb_{3.8}Mo_{1.2}O_{20.1}$. For example, $Ba_7Nb_{3.8}Mo_{1.2}O_{20.1}$ and $Ba_3MoNbO_{8.5}$ are similar hexagonal perovskite-related oxides, consist of the same elements, and have similar diffusion pathways of oxide ions through apical and interstitial oxygen sites in the c′ layers (Fig. 3.28b, c). However, the $E_a$ of $Ba_7Nb_{3.8}Mo_{1.2}O_{20.1}$ (0.25(4)–0.41(4) eV from 306 to 593 °C) is significantly lower than that of $Ba_3MoNbO_{8.5}$ (0.740(15) eV from 505–604 °C) [52]. Low $E_a$ values from 0.222(7) eV to 0.30(3) eV in dry air have been reported also in $Ba_7Nb_{4-x}Mo_{1+x}O_{20+x/2}$ ($x = 0, 0.1$) [13, 15]. We discuss the structural origin of the low $E_a$ in $Ba_7Nb_{4-x}Mo_{1+x}O_{20+x/2}$ in terms of the (Nb/Mo)–Ba distance at RT, because the oxide-ion diffusion pathways in the c′ layers are sandwiched between the Nb/Mo and Ba cations along the *c* axis (Fig. 3.31a), and hence the distance between the two cations (Nb/Mo) and Ba are regarded as a measure of an effective bottleneck for the oxide-ion migration. The (Nb/Mo)–Ba2 distances in the $Ba_7Nb_4MoO_{20}$-based materials (4.374(9)–4.397(3) Å) are longer than that in $Ba_3MoNbO_{8.5}$ (4.074(3) Å), which negatively correlates with the observed $E_a$ values (Fig. 3.31b) We have found that this negative correlation is applicable not only to $Ba_7Nb_4MoO_{20}$-based

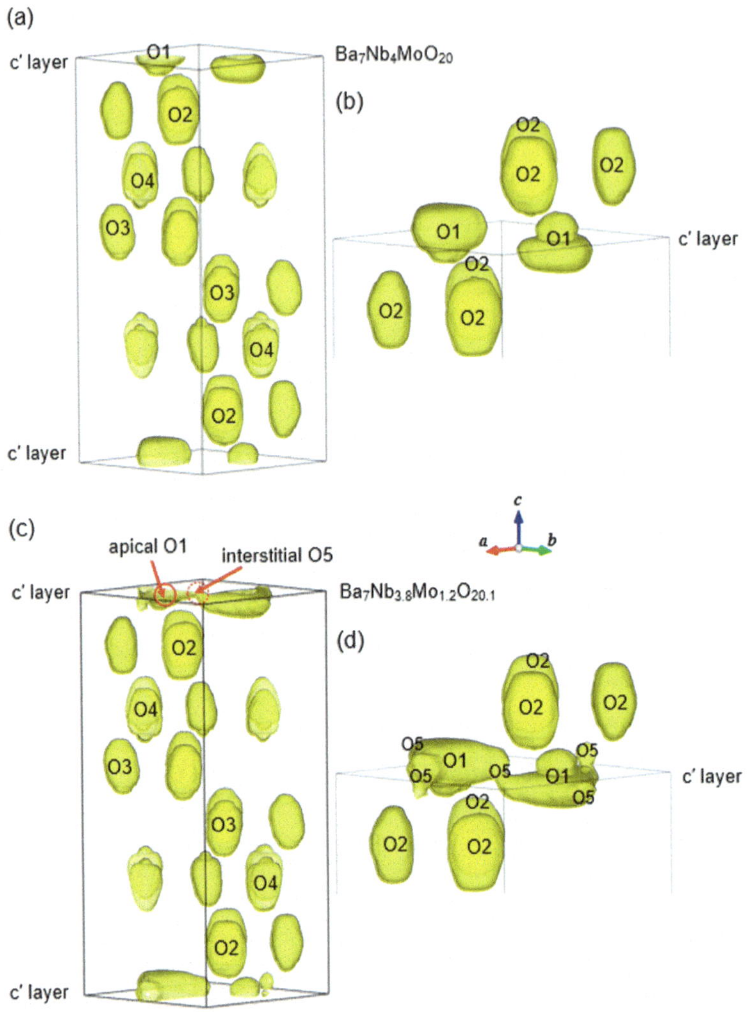


**Fig. 3.26** Time averaged probability-density distributions of oxide ions in $Ba_7Nb_{4-x}Mo_{1+x}O_{20+x/2}$ ((**a**, **b**) $x = 0$ and (**c**, **d**) $x = 0.2$), which were obtained by AIMD simulations at 1200 °C where the probability-density distributions in the 3 × 3 × 1 supercells were also averaged to 1 × 1 × 1. Yellow isosurfaces at 0.7 $Å^{-3}$ in the ranges of **a** $0 \le z \le 1$ ($Ba_7Nb_4MoO_{20}$), **b** $-0.2 \le z \le 0.2$ ($Ba_7Nb_4MoO_{20}$), **c** $0 \le z \le 1$ ($Ba_7Nb_{3.8}Mo_{1.2}O_{20.1}$), and **d** $-0.2 \le z \le 0.2$ ($Ba_7Nb_{3.8}Mo_{1.2}O_{20.1}$). In $Ba_7Nb_{3.8}Mo_{1.2}O_{20.1}$ and $Ba_7Nb_4MoO_{20}$, the spatial distribution of O1 is not connected to O2, indicating that rotational motions among the O1 and the three O2 atoms are quite rare. Thus, the three O2 atoms are immobile, while the apical O1 and interstitial O5 atoms are mobile. Copyright© 2023 Authors

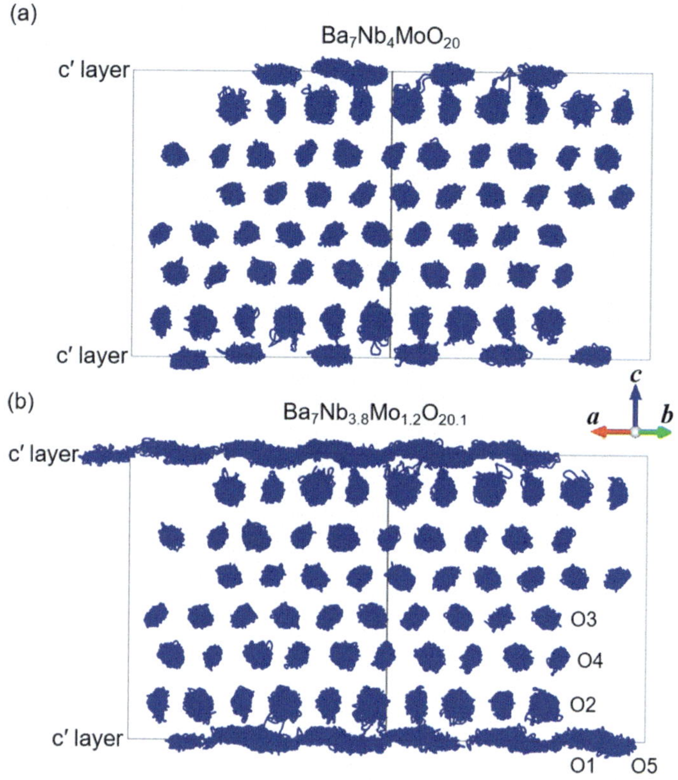


**Fig. 3.27** Trajectories of oxide ions in **a** $Ba_7Nb_4MoO_{20}$ and **b** $Ba_7Nb_{3.8}Mo_{1.2}O_{20.1}$ obtained by the AIMD simulations at 1200 °C. O*i* denotes the oxygen O*i* site in the average structure ($i = 1, 2, 3, 4, 5$; Tables 3.2 and 3.3). Copyright© 2023 Authors

oxides and $Ba_3MoNbO_{8.5}$ but also to related oxides $Ba_3(M, M')_2O_{8.5}$ ($M$ = Mo, W; $M'$ = V, Nb). As shown in Fig. 3.31b, $E_a$ decreases with an increase of the corresponding ($M/M'$)–Ba length, from which we conclude that the longer ($M/M'$)–Ba lengths in $Ba_7Nb_4MoO_{20}$-based materials reduce the activation energy for oxide-ion conductivity. Why are the $M/M'$–Ba distances in $Ba_7Nb_4MoO_{20}$-based oxides longer than those in $Ba_3(M, M')_2O_{8.5}$? To answer this question, the Madelung energies of $Ba_7Nb_{3.8}Mo_{1.2}O_{20.1}$ and $Ba_3MoNbO_{8.5}$ [70] were calculated as a function of the (Nb/Mo)–Ba distance (Fig. 3.33). $Ba_7Nb_{3.8}Mo_{1.2}O_{20.1}$ exhibits a minimum energy at a (Nb/Mo)–Ba distance of 4.60 Å, while $Ba_3MoNbO_{8.5}$ has a minimum energy at the shorter distance of 4.31 Å. The oxygen-deficient c′ layer $BaO_{2.1}$ in $Ba_7Nb_{3.8}Mo_{1.2}O_{20.1}$ (–2.2*e*) is less negatively charged than $BaO_{2.5}$ in $Ba_3MoNbO_{8.5}$ (–2.72*e*), thus, the (Nb/Mo)–Ba2 distance in $Ba_7Nb_{3.8}Mo_{1.2}O_{20.1}$ is longer than that in $Ba_3MoNbO_{8.5}$.

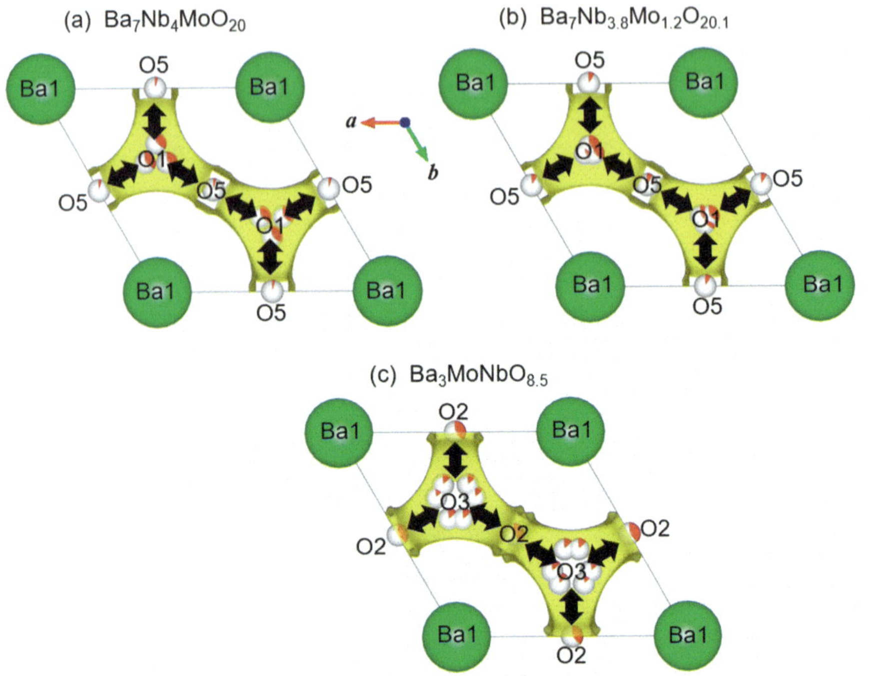


**Fig. 3.28** Bond-valence-based energy (BVE) landscapes and crystal structures on the *ab* plane at $z = 0$ of **a** $Ba_7Nb_4MoO_{20}$, **b** $Ba_7Nb_{3.8}Mo_{1.2}O_{20.1}$ and **c** $Ba_3MoNbO_{8.5}$ at 800 °C. Yellow isosurface levels: **a**, **b** 2.0 eV and **c** 1.8 eV. These BVE landscapes suggest –O1–O5– oxide-ion diffusion paths in $Ba_7Nb_4MoO_{20}$ and $Ba_7Nb_{3.8}Mo_{1.2}O_{20.1}$ and –O2–O3– pathways in $Ba_3MoNbO_{8.5}$. The diffusion pathways in $Ba_7Nb_4MoO_{20}$, $Ba_7Nb_{3.8}Mo_{1.2}O_{20.1}$ and $Ba_3MoNbO_{8.5}$ are similar, because the oxide ions two-dimensionally diffuse via the O1 (O3) and O5 (O2) sites along the $[1\bar{1}0]$, [120], and $[\bar{2}\bar{1}0]$ directions. Crystal structure and BVE landscape in panel (c) were obtained using the crystal data in [70]. Copyright© 2023 Authors

## 3.3.5 *Mechanisms of Hydration and Proton Conduction*

### 3.3.5.1 Crystal Structure Analyses of Hydrated (Deuterated) $Ba_7Nb_{3.8}Mo_{1.2}O_{20.1}$ (= $Ba_7Nb_{3.8}Mo_{1.2}O_{20.1}$·0.25(4) $D_2O$) and Dry $Ba_7Nb_{3.8}Mo_{1.2}O_{20.1}$

To investigate the occupancy and position of hydrogen and oxygen atoms, the crystal structures of hydrated (deuterated) $Ba_7Nb_{3.8}Mo_{1.2}O_{20.1}$ (= $Ba_7Nb_{3.8}Mo_{1.2}O_{20.1}$·0.25(4) $D_2O$) and dry $Ba_7Nb_{3.8}Mo_{1.2}O_{20.1}$ were analyzed using neutron-diffraction data at a low temperature 5 K where the thermal motions are suppressed. Rietveld analyses of these neutron diffraction data were successfully performed by the $P\bar{3}m1$ 7H polytype (Figs. 3.34 and 3.35). BVS values were reasonable (Tables 3.6 and 3.7), validating the refined crystal structures of hydrated and dry $Ba_7Nb_{3.8}Mo_{1.2}O_{20.1}$. The occupancy factors of O1 and O5 atoms were *g*(O;

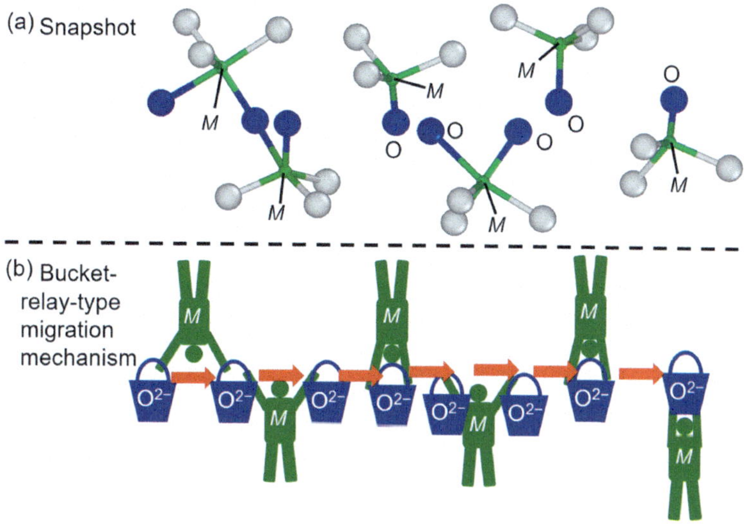


**Fig. 3.29** Snapshot (**a**) and corresponding schematic representation (**b**) of "bucket-relay"-type migration mechanism. In panel a, the blue spheres stand for the mobile apical and interstitial oxide ions in the oxide-ion conducting c′ layer. The grey and green spheres denote the relatively immobile O2 atoms forming the trigonal planes and *M* (= Nb/Mo2) cations, respectively. People and buckets in panel b correspond to *M* cations and mobile oxide ions, respectively. Copyright© 2023 Authors

O1) = 1/3 and $g$(O; O5) = 0.044(9), respectively, in dry sample (Table 3.7), while they were $g$(O; O1) = 1/3 and $g$(O; O5) = 0.117(13) in hydrated sample (Table 3.6). Thus, the oxygen content of hydrated sample is higher than that of dry sample, due to the incorporation of water into the c′ layer. In dry sample, no hydrogen atoms were observed (See the details in the footnote of Table 3.7), while in hydrated sample, the refined occupancy of D atom $g$(D; D) = 0.042(7) indicates the presence of hydrogen atoms. The hydrogen atoms are coordinated with the lattice O1 and interstitial O5 atoms in the c′ layer (Fig. 3.36). The O1–D and O5–D bond lengths were calculated to be 1.03(15) and 0.97(17) Å, respectively, which agreed with the O–H distance 1.00(11) Å estimated using the position of OH stretching Raman band (Fig. 3.37). The water content $y$ in hydrated $Ba_7Nb_{3.8}Mo_{1.2}O_{20.1} \cdot y\ D_2O$ was estimated to be 0.25(4) using the occupancy factors, which agreed with $y$ values 0.33 and 0.34 from TG and TG-MS measurements within three times of estimated standard deviation (Figs. 3.18 and 3.38). Since the excess oxygen atoms and hydrogen atoms due to the hydration are located in the c′ layer at 5 K, the water is incorporated forming the O–D (O–H) bonds in the c′ layer. Thus, the hydration of $Ba_7Nb_{3.8}Mo_{1.2}O_{20.1}$ occurs in the c′ layer, which is consistent with the conclusion on $Ba_7Nb_4MoO_{20} \cdot y\ D_2O$ in the literature [25, 80].

We also carried out Rietveld analyses of neutron diffraction data of $Ba_7Nb_{3.8}Mo_{1.2}O_{20.1} \cdot 0.25\ H_2O$ taken at 21 °C. The results indicated that the H atoms exist in the h layer (Table 3.5), which is consistent with the proton migration in the h layer observed in the AIMD simulations (See the details in the next section).

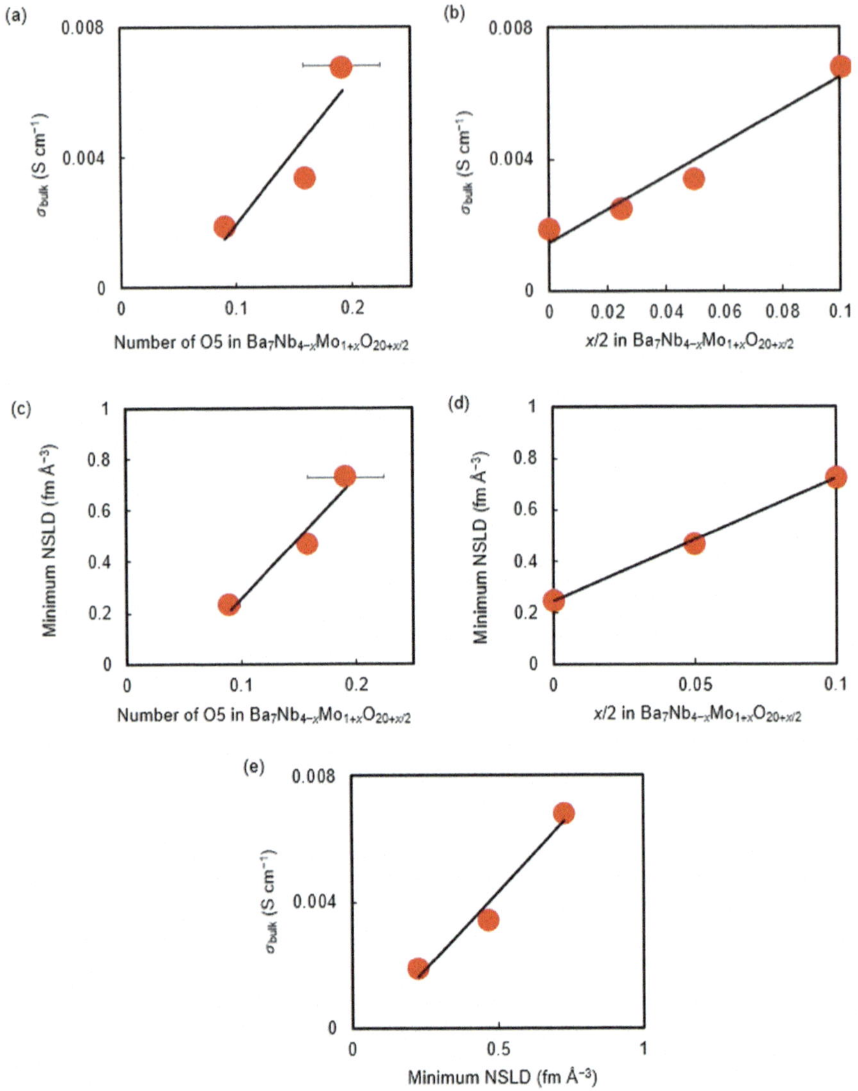


**Fig. 3.30** Variation of bulk conductivity at 500 °C with **a** the number of O5 atoms in a unit cell and **b** amount of excess oxygen atoms $x/2$ in $Ba_7Nb_{4-x}Mo_{1+x}O_{20+x/2}$ at 800 °C. Variation of the minimum NSLD between the O1 and O5 sites of $Ba_7Nb_{4-x}Mo_{1+x}O_{20+x/2}$ at 800 °C with **c** the number of O5 atoms in a unit cell and **d** amount of excess oxygen atoms $x/2$ in $Ba_7Nb_{4-x}Mo_{1+x}O_{20+x/2}$ at 800 °C. **e** Variation of bulk conductivity at 500 °C with the minimum NSLD between the O1 and O5 sites of $Ba_7Nb_{4-x}Mo_{1+x}O_{20+x/2}$ at 800 °C. Bulk conductivities at 500 °C of $Ba_7Nb_{4-x}Mo_{1+x}O_{20+x/2}$ ($x =$ 0, 0.05, 0.1) were taken from [13] and [15]. Minimum NSLD of $Ba_7Nb_{4-x}Mo_{1+x}O_{20+x/2}$ ($x = 0.1$) was obtained from [15]. Copyright© 2023 Authors

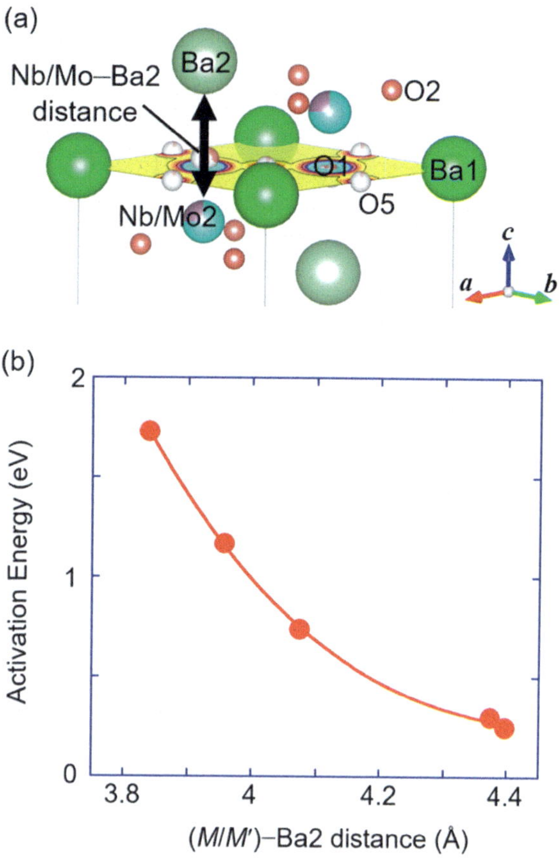


**Fig. 3.31** **a** NSLD distribution and crystal structures near the oxide-ion conducting c′ layer of $Ba_7Nb_{3.8}Mo_{1.2}O_{20.1}$ at 800 °C. The solid line with black arrows represents the Nb/Mo–Ba2 distances. **b** Variation of the activation energy for bulk conductivity in the temperature range from 500 to 600 °C with the *M*/*M*′–Ba2 distance at RT or 50 °C of hexagonal perovskite-related materials (*M* = Mo, W; *M*′ = V, Nb): $Ba_3WNbO_{8.5}$; [77] $Ba_3VWO_{8.5}$; [78] $Ba_3MoNbO_{8.5}$; [52, 70] $Ba_7Nb_{3.8}Mo_{1.2}O_{20.1}\cdot y\ H_2O$; (this work, Table 3.5) and $Ba_7Nb_4MoO_{20}$ [13]. Structural parameters of $Ba_7Nb_4MoO_{20}$ and $Ba_3MM'O_{8.5}$ (*M* = Mo, W *M*′ = Nb, V) are taken from the literature. Copyright© 2023 Authors

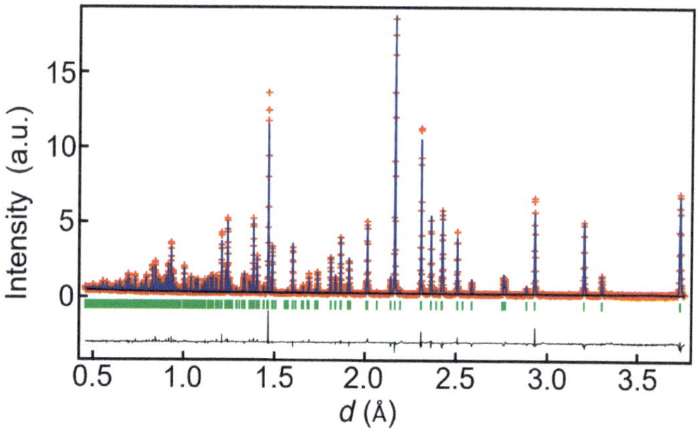


**Fig. 3.32** Rietveld patterns of as-prepared $Ba_7Nb_{3.8}Mo_{1.2}O_{20.1}\cdot y\ H_2O$ of neutron-diffraction data taken with the backscattering bank (*d* = 0.46–3.76 Å) of the SPICA diffractometer in vacuum at 21 °C. Red crosses, dark blue lines, green bars, and black lines represent observed intensities, calculated intensities, peak positions, and difference plots, respectively. Copyright© 2023 Authors

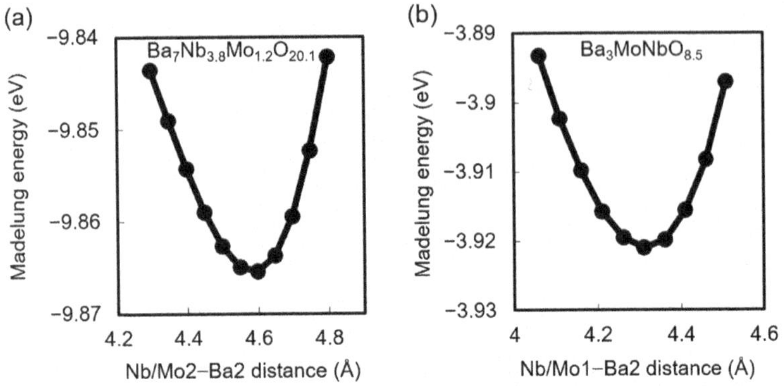


**Fig. 3.33** Madelung energy as a function of the Nb/Mo–Ba2 distance in **a** $Ba_7Nb_{3.8}Mo_{1.2}O_{20.1}$ and **b** $Ba_3MoNbO_{8.5}$ [70] at 800 °C. The Madelung energy for the unit cell was calculated using a Fourier method and refined crystal parameters with a program MADEL embedded in *VESTA 3* program [40, 79]. Closed black circles denote the calculated data. Crystal parameters of $Ba_7Nb_{3.8}Mo_{1.2}O_{20.1}$ (Table 3.2) and $Ba_3MoNbO_{8.5}$ [70] were used where the atomic coordinates of Nb/Mo and Ba2 atoms were changed depending on the Nb/Mo–Ba2 distance. In the calculations with MADEL, the RADIUS (radius of an ionic sphere) and REGION (Reciprocal-space range within which Fourier coefficients are summed up) parameters [79] were set to be 1.5 Å and 3.0 $Å^{-1}$, respectively. Copyright© 2023 Authors

#### 3.3.5.2 Proton Migration Investigated by Ab Initio Molecular Dynamics (AIMD) Simulations of $Ba_7Nb_{3.8}Mo_{1.2}O_{20.1}$·0.1 $H_2O$ and $Ba_7Nb_4MoO_{20}$·0.1 $H_2O$

To investigate the proton diffusion mechanism, four AIMD simulations of [1], [2], [3] and [4] (Table 3.8) were performed using 3 × 3 × 1 supercells $Ba_{63}Nb_{34}Mo_{11}O_{182}H_2$ ~ $(Ba_7Nb_{3.8}Mo_{1.2}O_{20.1}\cdot 0.1\ H_2O)_9$ and $Ba_{63}Nb_{36}Mo_9O_{181}H_2$ ~ $(Ba_7Nb_4MoO_{20}\cdot 0.1\ H_2O)_9$. Here, the water content $y$ in $Ba_7Nb_{4-x}Mo_{1+x}O_{20+x/2}\cdot y\ H_2O$ ($x = 0$ and 0.2) was set to 0.1, because the $y$ value in $Ba_7Nb_{3.8}Mo_{1.2}O_{20.1}\cdot y\ H_2O$ is approximately 0.1 at around 500 °C where its proton conductivity is relatively high (Fig. 3.38).

Since the refined structure of hydrated $Ba_7Nb_{3.8}Mo_{1.2}O_{20.1}\cdot y\ H_2O$ at 5 K indicated that the hydrogen and excess oxygen atoms exist in the c′ layer, these atoms were put in the c′ layer of the initial structures (0 ps) for the AIMD simulation of $(Ba_7Nb_{3.8}Mo_{1.2}O_{20.1}\cdot 0.1\ H_2O)_9$ [1]. In the AIMD simulation [1], a hydrogen atom H1 is often coordinated with O5 (73%) rather than O1 atom (27%) (Fig. 3.39(b)), although the number of O5 atom (30%) is much smaller than that of O1 atom (70%). For example, the proton H1 is coordinated with the corner-sharing O5 atom of $M_2O_9$ dimer (Fig. 3.40). These results suggest that protons in the c′ layer prefer the positions near the corner-sharing O5 atom of the $M_2O_9$ dimer. Therefore, it is likely that the $M_2O_9$ dimer has an important role in the proton conduction in the c′ layer. The other proton H1 goes back and forth between the c′ to h layers from 0 to 30 ps (Figs. 3.40a and 3.41a–d; See the detailed explanation in the caption of Fig. 3.41). In the hexagonal perovskite-related oxide $Ba_2ScAlO_5$, protons similarly go back and forth between

**Table 3.5** Refined crystallographic parameters and reliability factors in the Rietveld analysis of the neutron-diffraction data of as-prepared $Ba_7Nb_{3.8}Mo_{1.2}O_{20.1}\cdot y\ H_2O$ taken with SPICA at 21 °C (Rietveld patterns are shown in Fig. 3.32)

| Site $s$ Atom label | Atom $X$ | Wyckoff position | $g(X; s)$ | $x$ | $y$ | $z$ | $U_{iso}$ (Å$^2$) | BVS (vu) |
|---|---|---|---|---|---|---|---|---|
| Ba1 | Ba | 1*a* | 1 | 0 | 0 | 0 | 0.00663(18) | 2.04 |
| Ba2 | Ba | 2*d* | 1 | 1/3 | 2/3 | 0.82725(12) | 0.00663(18) | 2.19 |
| Ba3 | Ba | 2*d* | 1 | 1 | 2/3 | 0.57448(13) | 0.00663(18) | 2.29 |
| Ba4 | Ba | 2*c* | 1 | 0 | 0 | 0.28013(13) | 0.00663(18) | 1.92 |
| Nb/Mo1 | $Nb_{0.76}Mo_{0.24}$ | 1*b* | 1 | 0 | 0 | 1/2 | 0.00464(13) | 4.28 |
| Nb/Mo2 | $Nb_{0.76}Mo_{0.24}$ | 2*d* | 1 | 1/3 | 2/3 | 0.09321(9) | 0.00464(13) | 4.74 |
| Nb/Mo3 | $Nb_{0.76}Mo_{0.24}$ | 2*d* | 1 | 1/3 | 2/3 | 0.34987(9) | 0.00464(13) | 4.43 |
| O1 | O | 6*i* | 1/3 | 0.3611(6) | 0.7222(13) | − 0.01349(17) | 0.0358(15) | 1.75 |
| O2 | O | 6*i* | 1 | 0.16723(19) | − 0.16723(19) | 0.13183(7) | 0.0128(2) | 1.89 |
| O3 | O | 6*i* | 1 | 0.16434(19) | − 0.16434(19) | 0.43136(7) | 0.00552(17) | 1.88 |
| O4 | O | 6*i* | 1 | 0.49385(13) | − 0.49385(13) | 0.29439(6) | 0.0058(2) | 1.88 |

(continued)

**Table 3.5** (continued)

| Site *s* Atom label | Atom *X* | Wyckoff position | $g(X; s)$ | $x$ | $y$ | $z$ | $U_{iso}$ (Å$^2$) | BVS (vu) |
|---|---|---|---|---|---|---|---|---|
| O5 | O | 3*e* | 0.117 | 1/2 | 0 | 0 | 0.0358(15) | 1.29 |
| H | H | 12*j* | 0.042 | 0.0727 | − 0.0613 | 0.1324 | 0.0358(15) | 0.87 |

Bravais lattice: hexagonal: Crystal system: trigonal. Space group: $P\overline{3}m1$. Lattice parameters: $a = b = 5.863739(9)$ Å, $c = 16.53411(5)$ Å. Number of formula per unit cell: $Z = 1$
$g(X; s)$: Occupancy factor of $X$ atom at the $s$ site. Linear constraints: $g$(Ba; Ba1) = $g$(Ba; Ba2) = $g$(Ba; Ba3) = $g$(Ba; Ba4) = $g$($Nb_{0.76}Mo_{0.24}$; Nb/Mo1) = $g$($Nb_{0.76}Mo_{0.24}$; Nb/Mo2) = $g$($Nb_{0.76}Mo_{0.24}$; Nb/Mo3) = $g$(O; O2) = $g$(O; O3) = $g$(O; O4) = 1
$U_{iso}(Xn)$: Isotropic atomic displacement parameter of $X$ atom at the $Xn$ site. Linear constraints: $U_{iso}$(Ba1) = $U_{iso}$(Ba2) = $U_{iso}$(Ba3) = $U_{iso}$(Ba4), $U_{iso}$(Nb/Mo1) = $U_{iso}$(Nb/Mo2) = $U_{iso}$(Nb/Mo3), $U_{iso}$(O1) = $U_{iso}$(O5) = $U_{iso}$(H)
Reliability factors: $R_{wp} = 6.03\%$, $R_p = 5.22\%$, $R_B = 6.23\%$, $R_F = 5.20\%$
The values of occupancy factors $g$(O; O1), $g$(O; O5) and $g$(H; H) in this $Ba_7Nb_{3.8}Mo_{1.2}O_{20.1}\cdot y\,H_2O$ sample are fixed to the values for $Ba_7Nb_{3.8}Mo_{1.2}O_{20.1}\cdot 0.25\,D_2O$ (Table S6). The atomic coordinates of H atoms were difficult to be refined and fixed to the values which were found after several trials
BVS: bond valence sum. BVSs were calculated using the bond-valence parameters reported by Gagné and Hawthorne [68]. Low BVS value at the interstitial O5 site 1.29 indicates that the O5 oxide ion is underbonded. Such underbonding at the interstitial site has been observed also in other compositions such as $Ba_7Nb_{3.9}Mo_{1.1}O_{20.05}$ [15]

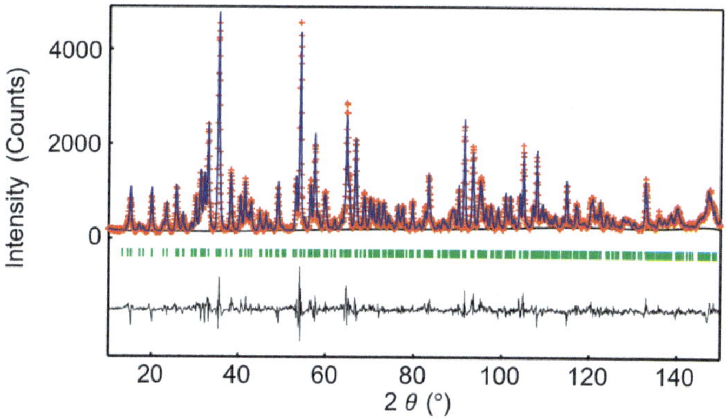


**Fig. 3.34** Rietveld pattern of in situ neutron-diffraction data of $Ba_7Nb_{3.8}Mo_{1.2}O_{20.1}$·0.25(4) $D_2O$ measured at 5 K with HERMES diffractometer. Red crosses, dark blue lines, green bars, and gray lines represent observed intensities, calculated intensities, peak positions, and difference plots, respectively. Copyright© 2023 Authors

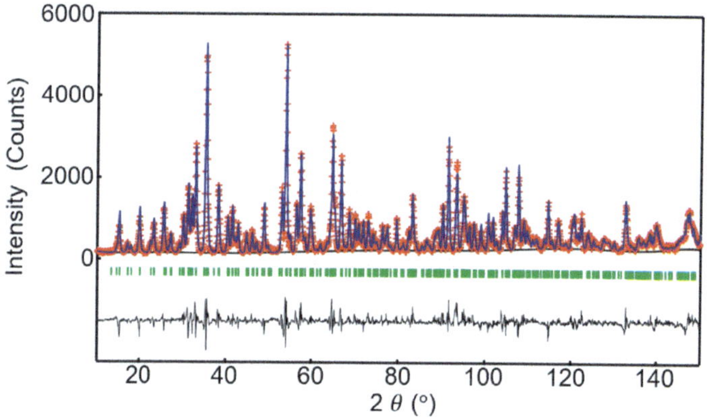


**Fig. 3.35** Rietveld pattern of in situ neutron-diffraction data of dry $Ba_7Nb_{3.8}Mo_{1.2}O_{20.1}$ measured at 5 K with HERMES diffractometer. Red crosses, dark blue lines, green bars, and gray lines represent observed intensities, calculated intensities, peak positions, and difference plots, respectively. Copyright© 2023 Authors

the h′ and c layers where the h′ layer represents the oxygen-deficient hexagonal close-packed layer [17]. In the AIMD simulation [1] of $(Ba_7Nb_{3.8}Mo_{1.2}O_{20.1}·0.1\ H_2O)_9$, the proton H1 migrates largely not only in the c′ layer (13–25 ps; Fig. 3.41d) but also in the h layer (25–46 ps; Fig. 3.41e). Both H1 and H2 protons migrate from the c′ to h layer (Figs. 3.40, 3.41a, e) in the AIMD simulation [1] of $(Ba_7Nb_{3.8}Mo_{1.2}O_{20.1}·0.1\ H_2O)_9$. We also observed proton migration from c′ to h layer in AIMD simulation [2] of mother material $(Ba_7Nb_4MoO_{20}·0.1\ H_2O)_9$ (Fig. 3.42).

Figure 3.41e shows that the proton H1 migrates largely in the h layers in the AIMD simulation [1] of $(Ba_7Nb_{3.8}Mo_{1.2}O_{20.1}·0.1\ H_2O)_9$. Thus, we have also performed the

**Table 3.6** Refined crystallographic parameters and reliability factors in the Rietveld analysis of the neutron-diffraction data of $Ba_7Nb_{3.8}Mo_{1.2}O_{20.1}$·0.25(4) $D_2O$ taken at 5 K with HERMES diffractometer

| Site *s* Atom label | Atom *X* | Wyckoff position | $g(X; s)$ | *x* | *y* | *z* | $U_{iso}$ (Å$^2$) | BVS (vu) |
|---|---|---|---|---|---|---|---|---|
| Ba1 | Ba | 1*a* | 1 | 0 | 0 | 0 | 0.0074(7) | 1.97 |
| Ba2 | Ba | 2*d* | 1 | 1/3 | 2/3 | 0.8282(5) | 0.0074(7) | 2.29 |
| Ba3 | Ba | 2*d* | 1 | 1 | 2/3 | 0.5749(5) | 0.0074(7) | 2.31 |
| Ba4 | Ba | 2*c* | 1 | 0 | 0 | 0.2820(5) | 0.0074(7) | 1.94 |
| Nb/Mo1 | $Nb_{0.76}Mo_{0.24}$ | 1*b* | 1 | 0 | 0 | 1/2 | 0.0080(5) | 4.37 |
| Nb/Mo2 | $Nb_{0.76}Mo_{0.24}$ | 2*d* | 1 | 1/3 | 2/3 | 0.0934(4) | 0.0080(5) | 4.69 |
| Nb/Mo3 | $Nb_{0.76}Mo_{0.24}$ | 2*d* | 1 | 1/3 | 2/3 | 0.3502(4) | 0.0080(5) | 4.35 |
| O1 | O | 6*i* | 1/3 | 0.353(6) | 0.705(11) | −0.0168(9) | 0.059(9) | 1.84 |
| O2 | O | 6*i* | 1 | 0.1683(8) | −0.1683(8) | 0.1328(3) | 0.0130(8) | 1.84 |
| O3 | O | 6*i* | 1 | 0.1645(7) | −0.1645(7) | 0.4323(3) | 0.0078(8) | 1.91 |
| O4 | O | 6*i* | 1 | 0.4945(6) | −0.4945(6) | 0.2935(2) | 0.0115(9) | 1.87 |
| O5 | O | 3*e* | 0.117(13) | 1/2 | 0 | 0 | 0.059(9) | 1.45 |
| D | H | 12*j* | 0.042(7) | 0.346(2) | 0.475(2) | −0.032(7) | 0.059(9) | 0.70 |

Bravais lattice: hexagonal: Crystal system: trigonal. Space group: $P\bar{3}m1$. Lattice parameters: $a = b = 5.85577(18)$ Å, $c = 16.5144(4)$ Å. Number of formula per unit cell: $Z = 1$

$g(X; s)$: Occupancy factor of *X* atom at the *s* site. Linear constraints: $g$(Ba; Ba1) = $g$(Ba; Ba2) = $g$(Ba; Ba3) = $g$(Ba; Ba4) = $g$($Nb_{0.76}Mo_{0.24}$; Nb/Mo1) = $g$($Nb_{0.76}Mo_{0.24}$; Nb/Mo2) = $g$($Nb_{0.76}Mo_{0.24}$; Nb/Mo3) = $g$(O; O2) = $g$(O; O3) = $g$(O; O4) = 1, $g$(O; O5) = 0.033 + 2 $g$(D; H). In a preliminary analysis, the occupancy factors of O5 and D were refined independently and the refined vaules gave the relation $g$(O; O5) = 0.033 + 2 $g$(D; H) within three estimated standard deviations. Thus, we applied this linear constraint in the final refinement

$U_{iso}(Xn)$: Isotropic atomic displacement parameter of *X* atom at the *Xn* site. Linear constraints: $U_{iso}$(Ba1) = $U_{iso}$(Ba2) = $U_{iso}$(Ba3) = $U_{iso}$(Ba4), $U_{iso}$(Nb/Mo1) = $U_{iso}$(Nb/Mo2) = $U_{iso}$(Nb/Mo3), $U_{iso}$(O1) = $U_{iso}$(O5) = $U_{iso}$(D)

Reliability factors: $R_{wp} = 11.93\%$, $R_p = 9.40\%$, $R_B = 2.83\%$, $R_F = 1.43\%$

BVS: bond valence sum. BVSs were calculated using the bond-valence parameters reported by Gagné and Hawthorne [68]. Low BVS value at the interstitial O5 site 1.45 indicates that the O5 oxide ion is underbonded. Such underbonding at the interstitial site has been observed also in other compositions such as $Ba_7Nb_{3.9}Mo_{1.1}O_{20.05}$ [15]

**Table 3.7** Refined crystallographic parameters and reliability factors in the Rietveld analysis of the neutron-diffraction data of dry $Ba_7Nb_{3.8}Mo_{1.2}O_{20.1}$ taken at 5 K with HERMES diffractometer

| Site *s* Atom label | Atom *X* | Wyckoff position | $g(X; s)$ | $x$ | $y$ | $z$ | $U_{iso}$ (Å$^2$) | BVS (vu) |
|---|---|---|---|---|---|---|---|---|
| Ba1 | Ba | 1*a* | 1 | 0 | 0 | 0 | 0.0067(6) | 1.91 |
| Ba2 | Ba | 2*d* | 1 | 1/3 | 2/3 | 0.8256(4) | 0.0067(6) | 2.19 |
| Ba3 | Ba | 2*d* | 1 | 1 | 2/3 | 0.5740(5) | 0.0067(6) | 2.33 |
| Ba4 | Ba | 2*c* | 1 | 0 | 0 | 0.2832(5) | 0.0067(6) | 1.94 |
| Nb/Mo1 | $Nb_{0.76}Mo_{0.24}$ | 1*b* | 1 | 0 | 0 | 1/2 | 0.0056(5) | 4.42 |
| Nb/Mo2 | $Nb_{0.76}Mo_{0.24}$ | 2*d* | 1 | 1/3 | 2/3 | 0.0944(4) | 0.0056(5) | 4.62 |
| Nb/Mo3 | $Nb_{0.76}Mo_{0.24}$ | 2*d* | 1 | 1/3 | 2/3 | 0.3491(3) | 0.0056(5) | 4.38 |
| O1 | O | 6*i* | 1/3 | 0.3589(16) | 0.718(3) | −0.0142(6) | 0.022(4) | 1.67 |
| O2 | O | 6*i* | 1 | 0.1677(8) | −0.1677(8) | 0.1319(2) | 0.0146(8) | 1.83 |
| O3 | O | 6*i* | 1 | 0.1632(7) | −0.1632(7) | 0.4319(3) | 0.0073(6) | 1.91 |
| O4 | O | 6*i* | 1 | 0.4953(6) | −0.4953(6) | 0.2952(2) | 0.0093(8) | 1.91 |
| O5 | O | 3*e* | 0.033 | 1/2 | 0 | 0 | 0.022(4) | 1.27 |

Bravais lattice: hexagonal: Crystal system: trigonal. Space group: $P\overline{3}m1$. Lattice parameters: $a = b = 5.85829(15)$ Å, $c = 16.5251(3)$ Å. Number of formula per unit cell: $Z = 1$

$g(X; s)$: Occupancy factor of $X$ atom at the $s$ site. Linear constraints: $g$(Ba; Ba1) = $g$(Ba; Ba2) = $g$(Ba; Ba3) = $g$(Ba; Ba4) = $g$($Nb_{0.76}Mo_{0.24}$; Nb/Mo1) = $g$($Nb_{0.76}Mo_{0.24}$; Nb/Mo2) = $g$($Nb_{0.76}Mo_{0.24}$; Nb/Mo3) = $g$(O; O2) = $g$(O; O3) = $g$(O; O4) = 1. In a preliminary analysis, the $g$(O; O1) and $g$(O; O5) were refined to be higher than 1 and lower than 0, respectively. In other analysis, the refined $g$(H; H) was minus, indicating no water incorporation. Therefore, the $g$(O; O1) and $g$(O; O5) were fixed in the final refinement

$U_{iso}(Xn)$: Isotropic atomic displacement parameter of $X$ atom at the $Xn$ site. Linear constraints: $U_{iso}$(Ba1) = $U_{iso}$(Ba2) = $U_{iso}$(Ba3) = $U_{iso}$(Ba4), $U_{iso}$(Nb/Mo1) = $U_{iso}$(Nb/Mo2) = $U_{iso}$(Nb/Mo3), $U_{iso}$(O1) = $U_{iso}$(O5)

Reliability factors: $R_{wp} = 12.43\%$, $R_p = 9.79\%$, $R_B = 4.60\%$, $R_F = 2.77\%$

BVS: bond valence sum. BVSs were calculated using the bond-valence parameters reported by Gagné and Hawthorne [68]. Low BVS value at the interstitial O5 site 1.27 indicates that the O5 oxide ion is underbonded. Such underbonding at the interstitial site has been observed also in other compositions such as $Ba_7Nb_{3.9}Mo_{1.1}O_{20.05}$ [15]

AIMD simulations [3] and [4], where two protons are placed in the h layers in the initial structures (0 ps). In the mother material ($Ba_7Nb_4MoO_{20}$·0.1 $H_2O$)$_9$, the MSD of the proton in the h layer in the AIMD simulation [3] is lower than that in c′ layer in the simulation [2] (Fig. 3.43), which is consistent with the conclusion in the literature [23]. In stark contrast, in ($Ba_7Nb_{3.8}Mo_{1.2}O_{20.1}$·0.1 $H_2O$)$_9$, the MSD

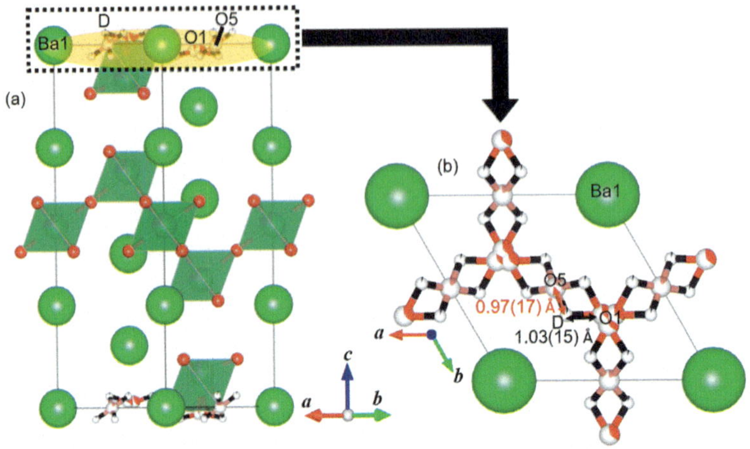


**Fig. 3.36** Refined crystal structure of wet (deuterated) $Ba_7Nb_{3.8}Mo_{1.2}O_{20.1}$·0.25(4) $D_2O$ at 5 K. **a** Structure viewed along the [110] direction ($0 \leq x, y, z \leq 1$), which are drawn with (Nb/Mo)–O polyhedra. **b** Structure of the c′ layer viewed along the $c$ axis ($-0.05 \leq z \leq 0.05$). Red/black O1–D and pink/black O5–D bonds are drawn. Copyright© 2023 Authors

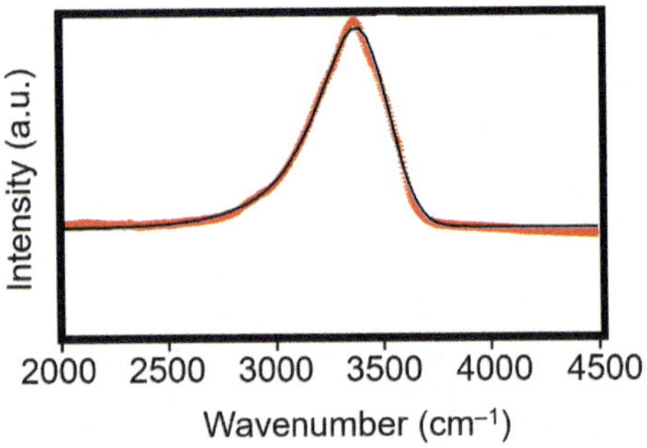


**Fig. 3.37** Raman spectrum of O–H stretching vibrations of as-prepared $Ba_7Nb_{3.8}Mo_{1.2}O_{20.1}\cdot y$ $H_2O$. Red cross marks and black line show experimental data and fitting curve, respectively. The O–H distance was calculated to be 1.00(11) Å using the empirical equation after Novak [81]. Copyright© 2023 Authors

of proton in the h layer in the AIMD simulation [4] is much higher than that in c′ layer in the AIMD simulation [1] (Fig. 3.44). Indeed, the trajectory of the proton indicates long-range proton migration in the h layer across the unit cells in the AIMD simulation [4] of $(Ba_7Nb_{3.8}Mo_{1.2}O_{20.1}\cdot 0.1\ H_2O)_9$ (Fig. 3.45). These results indicate that the proton conduction occurs mainly via the long-range proton migration in the h layer in $Ba_7Nb_{3.8}Mo_{1.2}O_{20.1}\cdot y$ $H_2O$. It was found that the MSD of protons in the h layer of $(Ba_7Nb_{3.8}Mo_{1.2}O_{20.1}\cdot 0.1\ H_2O)_9$ (AIMD simulation [4]) is higher than that in the c′ layer of $(Ba_7Nb_4MoO_{20}\cdot 0.1\ H_2O)_9$ (AIMD simulation [2]) (Fig. 3.46), which suggests the higher proton diffusivity of $(Ba_7Nb_{3.8}Mo_{1.2}O_{20.1}\cdot 0.1\ H_2O)_9$ than that

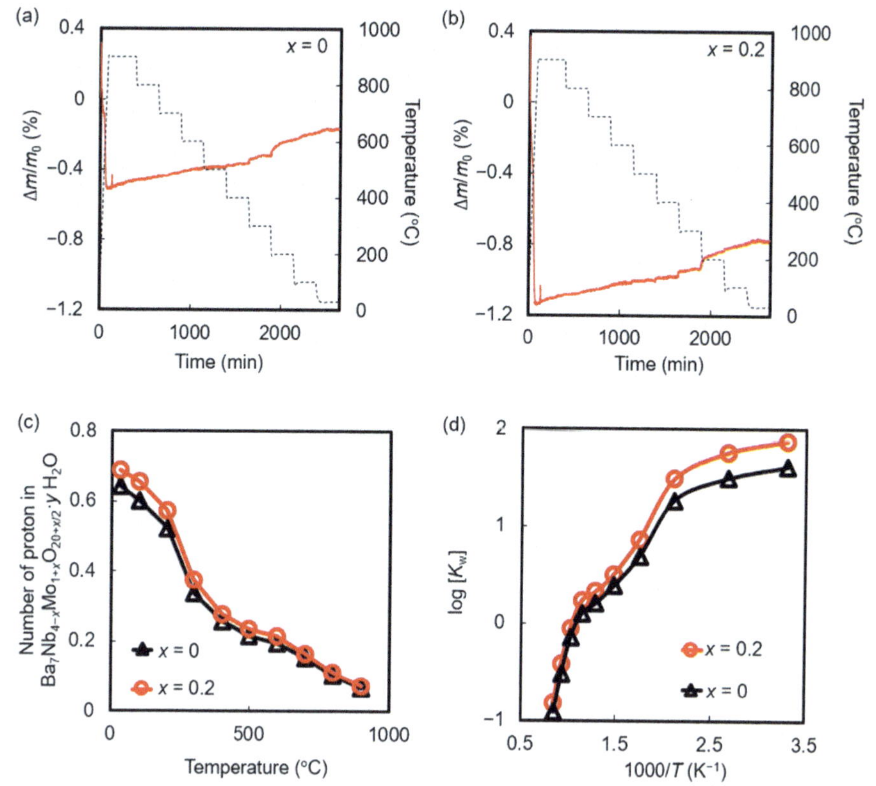


**Fig. 3.38** Thermogravimetric (TG) data of $Ba_7Nb_{4-x}Mo_{1+x}O_{20+x/2}\cdot y$ $H_2O$ by equilibrium isotherms for **a** $x = 0$ and **b** $x = 0.2$ on heating under dry air and cooling under wet air (water vapor partial pressure: $P(H_2O) = 0.021$ atm). **c** Temperature dependence of proton concentration (= number of protons) $2y$ in $Ba_7Nb_{4-x}Mo_{1+x}O_{20+x/2}\cdot y$ $H_2O$ ($x = 0, 0.2$). It should be noted that the proton concentration for $x = 0.2$ is a little higher than that for $x = 0$. **d** van't Hoff plot for the water incorporation in wet air ($P(H_2O) = 0.021$ atm). $K_w$ values of $Ba_7Nb_{4-x}Mo_{1+x}O_{20+x/2}\cdot y$ $H_2O$ ($x =$ 0, 0.2) were calculated by using the formula (3.1) in Sect. 3.3.6.3. Copyright© 2023 Authors

**Table 3.8** Four AIMD calculations of [1–4]

| AIMD No. | Composition of $Ba_7Nb_{4-x}Mo_{1+x}O_{20+x/2}\bullet 0.1$ $H_2O$ | Initial position of H1 atom at 0 fs | Initial position of H2 atom at 0 fs | |
|---|---|---|---|---|
| [1] | $x = 0.2$ | c′ | c′ | 3.39, 3.41 3.44, 3.46 (Black line) |
| [2] | $x = 0.0$ | c′ | c′ | 3.43 3.46 (Black line) |
| [3] | $x = 0.0$ | h | h | 3.43 (Red line) |
| [4] | $x = 0.2$ | h | h | 3.39, 3.41 |

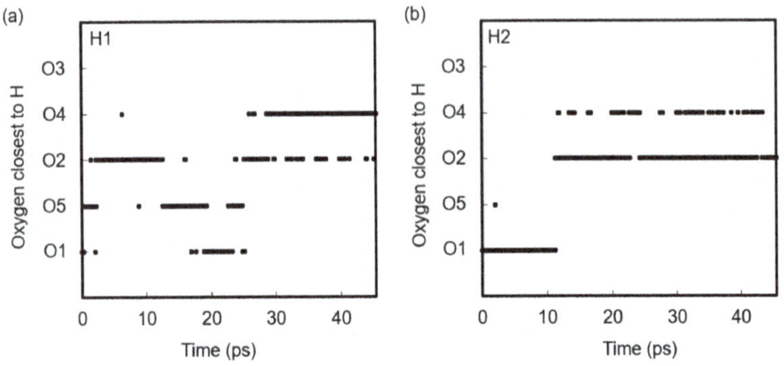


**Fig. 3.39** Oxygen O1–O5 atoms closest to **a** H1 and **b** H2 atoms in $Ba_7Nb_{3.8}Mo_{1.2}O_{20.1}{\cdot}0.1\ H_2O$ at 1200 °C in the AIMD simulation [1]. Here, O1 denotes $O_{ap}$ and $O_{ap}'$ atoms, while O5 stands for $O_{sh}$ atom. Oxygen atoms O1 and O5 exist in the c′ layer. The O2 and O4 atoms are located in different h layers, while the O3 atom exists in the c layer (Fig. 3.21). Copyright© 2023 Authors

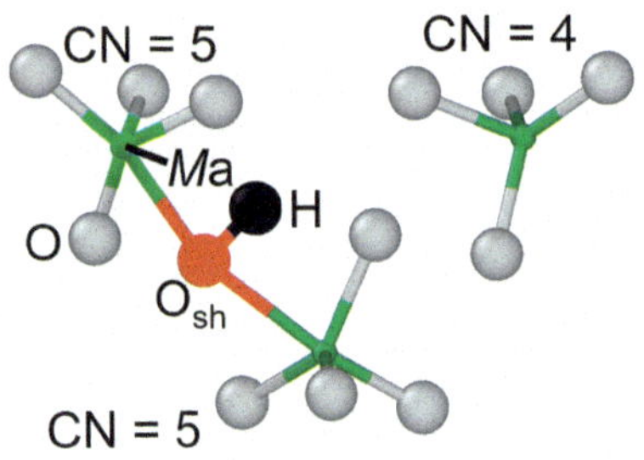


**Fig. 3.40** Proton migration process of $Ba_7Nb_{3.8}Mo_{1.2}O_{20.1}{\cdot}0.1H_2O$ in the c′ layer at 1200 °C in the AIMD simulation [1]. Red, gray, black, and green spheres represent corner-sharing oxygen atom $O_{sh}$, oxygen atom, hydrogen atom and Nb or Mo cations (*M*: Nb/Mo2), respectively. Copyright© 2023 Authors

of $(Ba_7Nb_4MoO_{20}{\cdot}0.1\ H_2O)_9$. Both higher diffusivity and concentration of protons would be the origins of higher proton conductivity in $Ba_7Nb_{3.8}Mo_{1.2}O_{20.1}{\cdot}y\ H_2O$, compared with the mother material $Ba_7Nb_4MoO_{20}{\cdot}y\ H_2O$ (Figs. 3.18b and 3.38).

### 3.3.6 Supplementary Note

#### 3.3.6.1 Kramers–Kronig (KK) Validation

Figure 3.47 show the KK residuals of experimental data of $Ba_7Nb_{3.8}Mo_{1.2}O_{20.1}$ in dry air and $Ba_7Nb_{3.8}Mo_{1.2}O_{20.1}{\cdot}y\ H_2O$ in wet air. The absolute values of the KK residual were lower than 0.6%.

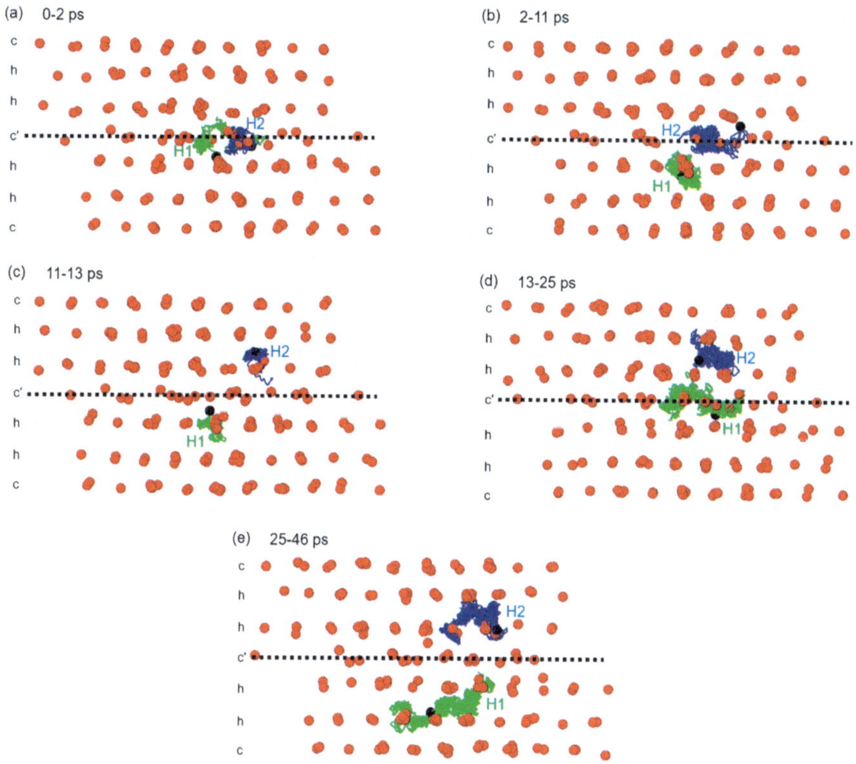


**Fig. 3.41** Trajectories of protons in $Ba_{63}Nb_{34}Mo_{11}O_{181}\cdot H_2O$ [~ $(Ba_7Nb_{3.8}Mo_{1.2}O_{20.1}\cdot 0.1\ H_2O)_9$] in the AIMD simulation [1] at 1200 °C during **a** 0–2 ps, **b** 2–11 ps, **c** 11–13 ps, **d** 13–25 ps, and **e** 25–46 ps. In the initial state (0 ps), both H1 and H2 atoms were put in the c′ layer. Green and blue lines stand for the trajectories of H1 and H2 atoms, respectively. A black dot lines denotes the c′ layer. Red spheres denote oxygen atoms. The H2 proton migrates from c′ to h layer at 11 ps (Figs. 3.39, 3.41b, c). The other proton H1 exists mainly near O5 ($O_{sh}$) atom in the c′ layer between 0 and 2 ps, and then migrates from c′ to h layer at 2 ps (Fig. 3.41a, b), returns to c′ layer at 13 ps (Fig. 3.41c, d), goes back and forth between the positions near the O5 and O1 atoms in the c′ layer between 13 and 25 ps, and the H1 atom moves from c′ to h layer again at 25 ps, and then goes back and forth between the positions near the O2 and O4 atoms in different h layers between 25 and 46 ps (Fig. 3.41d, e). Thus, the H1 goes back and forth between the c′ to h layers from 0 to 25 ps. 

Figure 3.48 shows the equivalent circuit (EC) residual plots of $Ba_7Nb_{3.8}Mo_{1.2}O_{20.1}$ in dry air and $Ba_7Nb_{3.8}Mo_{1.2}O_{20.1}\cdot y\ H_2O$ in wet air. The absolute values of the EC residual were lower than 0.1%.

These residual values validate the collected impedance data [82, 83].

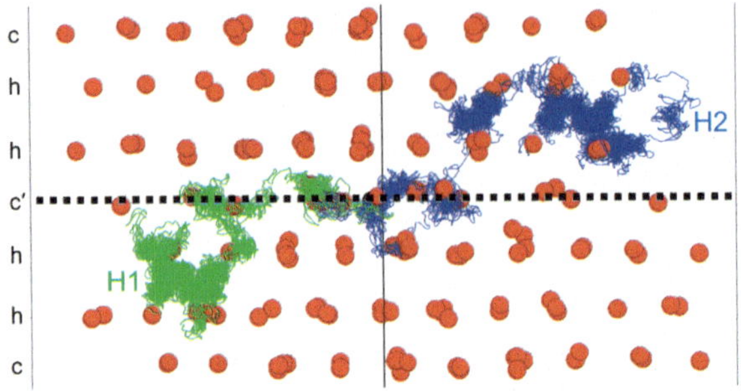


**Fig. 3.42** Trajectories of protons in $Ba_{63}Nb_{36}Mo_9O_{180}{\cdot}H_2O$ [~ $(Ba_7Nb_4MoO_{20}{\cdot}0.1\ H_2O)_9$] obtained by the AIMD simulation [2] at 1200 °C during 0–68 ps. Two H atoms were placed only in the c′ layer at 0 ps in the AIMD simulation [2]. Black line denotes the 3 × 3 × 1 supercell. A black dot lines denotes the c′ layer. Red spheres denote oxygen atoms. Copyright© 2023 Authors

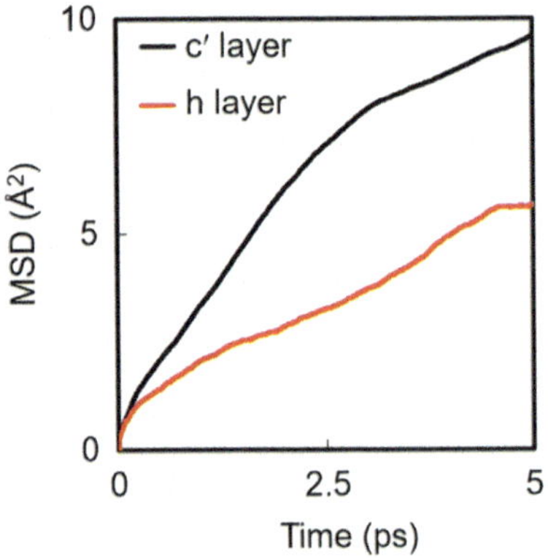


**Fig. 3.43** MSDs of H atoms of $Ba_{63}Nb_{36}Mo_9O_{180}{\cdot}H_2O$ [~ $(Ba_7Nb_4MoO_{20}{\cdot}0.1\ H_2O)_9$] in AIMD simulations [2] and [3] at 1200 °C. Two H atoms were placed only in the c′ layer at 0 ps in the AIMD simulation [2] (black curve), while two H atoms were placed only in the h layer at 0 ps in the AIMD simulations [3] (red curve). Copyright© 2023 Authors

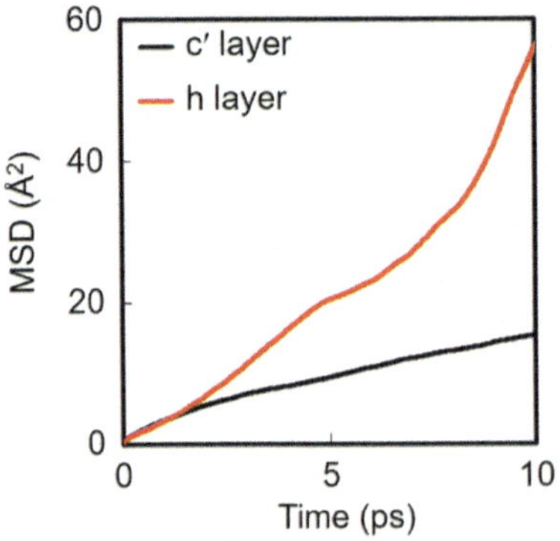


**Fig. 3.44** MSDs of H atoms of $Ba_{63}Nb_{34}Mo_{11}O_{181}{\cdot}H_2O$ [~ $(Ba_7Nb_{3.8}Mo_{1.2}O_{20.1}{\cdot}0.1\ H_2O)_9$] in AIMD simulations [1] and [4] at 1200 °C. Two H atoms were placed only in the c′ layer at 0 ps in the AIMD simulation [1] (black curve), while two H atoms were placed only in the h layer at 0 ps in the AIMD simulations [4] (red curve). Copyright© 2023 Authors

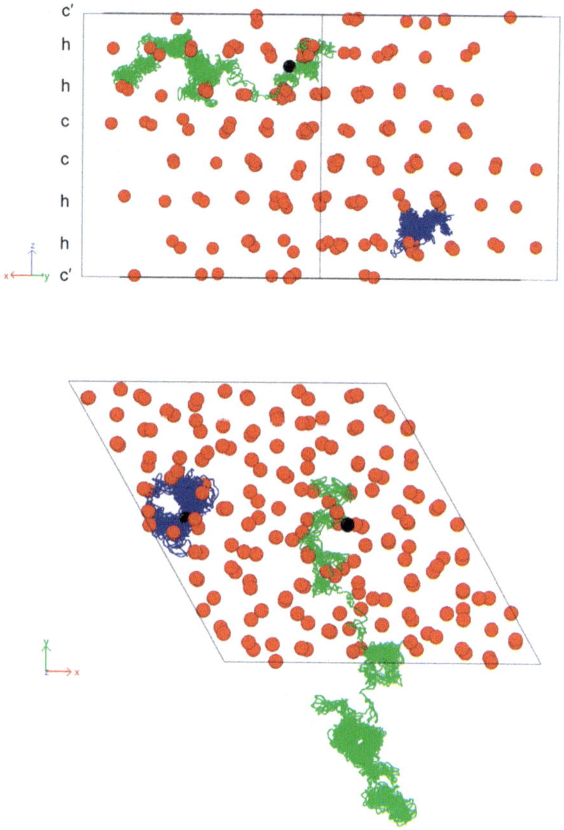


**Fig. 3.45** Trajectories of protons in $(Ba_7Nb_{3.8}Mo_{1.2}O_{20.1}\cdot 0.1\ H_2O)_9$ obtained by the AIMD simulation [4] at 1200 °C at 0–38 ps. Two H atoms are placed only in the h layer at 0 ps in the AIMD simulation [4]. Black line denotes the 3 × 3 × 1 supercell. A proton comes across 1 × 1 × 1 unit cells (green trajectory). Copyright© 2023 Authors

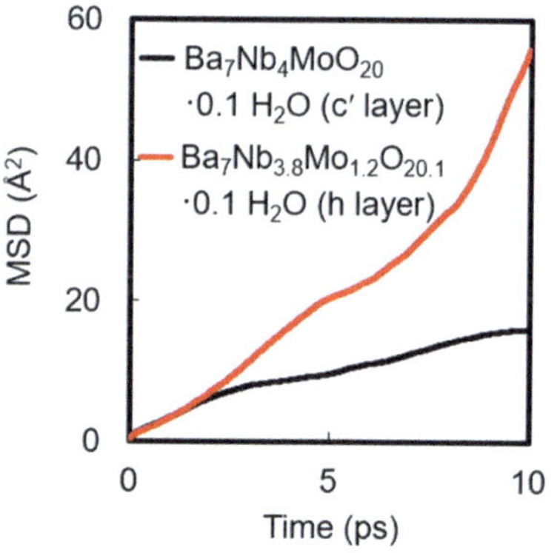


**Fig. 3.46** MSD of H atoms of $Ba_{63}Nb_{34}Mo_{11}O_{181}\cdot H_2O$ [~ $(Ba_7Nb_{3.8}Mo_{1.2}O_{20.1}\cdot 0.1\ H_2O)_9$] in AIMD simulations [4] at 1200 °C where two H atoms were put only in the h layer at 0 ps (red curve). MSD of H atoms of $Ba_{63}Nb_{36}Mo_9O_{180}\cdot H_2O$ [~ $(Ba_7Nb_4MoO_{20}\cdot 0.1\ H_2O)_9$] in AIMD simulation [2] at 1200 °C where two H atoms were placed only in the c′ layer at 0 ps (black curve). Copyright© 2023 Authors

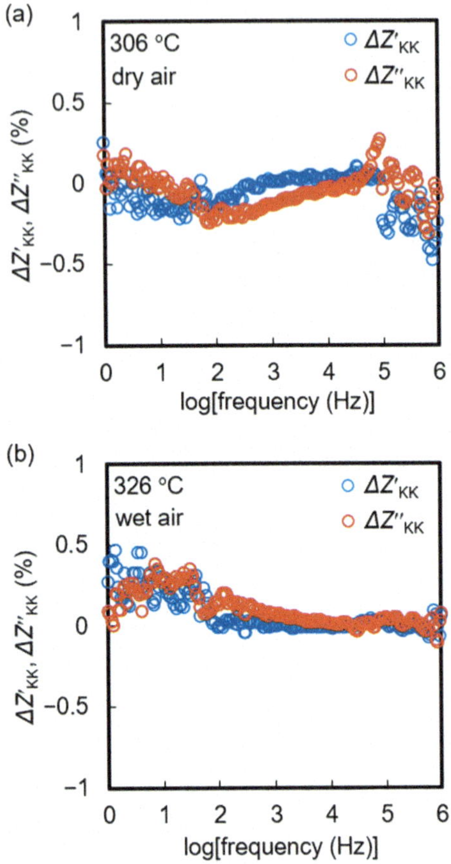


**Fig. 3.47** Residual plots of the measured impedance data of **a** $Ba_7Nb_{3.8}Mo_{1.2}O_{20.1}$ in dry air at 306 °C and **b** $Ba_7Nb_{3.8}Mo_{1.2}O_{20.1}\cdot yH_2O$ in wet air at 326 °C obtained by the Kramers–Kronig transformation. Copyright© 2023 Authors

#### 3.3.6.2 Experimental Support for the Presence of $M_2O_9$ Dimers

In the AIMD simulation, the $M$–$O_{sh}$ distance (1.91–2.35 Å) is longer than $M$–$O_{ap}$ distance (1.67–1.90 Å) and the $M$–$O_{ap}$' distance (1.74–1.92 Å). Here the $M$ stands for the Nb/Mo2 atom in the AIMD simulations, $O_{sh}$ denotes corner-sharing oxygen atom of $M_2O_9$ dimer, the $O_{ap}$ stands for the apical oxygen atom in $MO_4$ tetrahedron, and $O_{ap}$' denotes the apical oxygen atom in $MO_5$ polyhedron. The Nb/Mo2–O5 distance (2.30 Å) in the refined structure using neutron diffraction data is longer than the $M2$–$O_{ap}$ distance (1.67–1.90 Å) and the $M2$–$O_{ap}$' distance (1.74–1.92 Å), indicating that the O5 atom is not the apical oxygen atom in $M2O_n$ ($n = 4$ and 5). In contrast, the Nb/Mo2–O5 distance (2.30 Å) in the refined structure using neutron diffraction data is in agreement with the $M$–$O_{sh}$ distance (1.91–2.35 Å) from the AIMD simulations, thus, the interstitial O5 atom corresponds to the $O_{sh}$ atom. Since the $O_{sh}$ atom has a large spatial distribution around the O5 position, the Nb/Mo2–O5 distance (2.30 Å) is close to the longest $M$–$O_{sh}$ distance in the AIMD simulations (2.35 Å). The Nb/Mo2–O1 distance in the refined structure using neutron diffraction

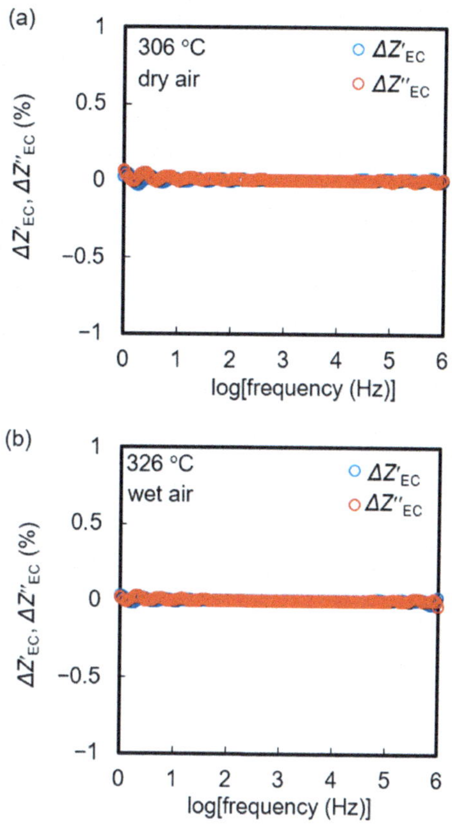


**Fig. 3.48** Residual plots of the equivalent circuit fitting of **a** $Ba_7Nb_{3.8}Mo_{1.2}O_{20.1}$ in dry air at 306 °C and **b** $Ba_7Nb_{3.8}Mo_{1.2}O_{20.1}\cdot y\,H_2O$ in wet air 326 °C. Copyright© 2023 Authors

data (1.79 Å) is in agreement with the $M$–$O_{ap}$ and $M$–$O_{ap}'$ distances (1.67–1.92 Å), therefore, the apical O1 atom in the average structure corresponds to the $O_{ap}$ and $O_{ap}'$ atoms in the AIMD simulations. Since O5 atom corresponds to the $O_{sh}$ atom, the presence of O5 atom in the Rietveld and MEM analyses experimentally supports the presence of $O_{sh}$ atom and $M_2O_9$ dimer.

#### 3.3.6.3 The Equilibrium Constant ($K_w$) of $Ba_7Nb_{4-x}Mo_xO_{20+x/2}\cdot y$ $H_2O$ ($x = 0, 0.2$)

The hydration of $Ba_7Nb_{4-x}Mo_xO_{20+x/2}\cdot y$ $H_2O$ occurs by the following defect equation.

$$H_2O(g) + V_{OV}^{\times} + O_{OL}^{\times} \leftrightarrow (O_{OV}H)' + (O_{OL}H)^{\cdot}$$

Here, OV and OL represent an interstitial O5 site and a lattice oxygen O1 site in the $c'$ layer, respectively. The equilibrium constant is

$$K_w = \frac{[(O_{OV}H)'][(O_{OL}H)^{\cdot}]}{P(H_2O)[V_{OV}^{\times}][O_{OL}^{\times}]}$$

The electroneutrality condition is

$$[(O_{OV}H)'] = [(O_{OL}H)^{\cdot}]$$

The conservations of the total numbers of vacancies and protons at the interstitial O5 is

$$[V_{OV}^{\times}] + [(O_{OV}H)'] = 1 - x/2$$

The conservations of the total numbers of oxygen atoms and protons at the lattice O1 is

$$[O_{OL}^{\times}] + [(O_{OL}H)^{\cdot}] = 2 + x/2$$

Therefore, the $K_w$ for the water incorporation reaction of $Ba_7Nb_{4-x}Mo_xO_{20+x/2} \cdot y$ $H_2O$ ($x = 0, 0.2$) can be written as follows.

$$\begin{aligned} K_w &= \frac{[(O_{OV}H)']^2}{P(H_2O)((1 - x/2) - [(O_{OV}H)'])((2 + x/2) - [(O_{OV}H)'])} \\ &= \exp[(-\Delta H^0 + T\Delta S^0)/RT] \end{aligned} \tag{3.1}$$

Here, the $\Delta H^0$, $\Delta S^0$ and $R$ are hydration enthalpy, hydration entropy and gas constant, respectively. The $\Delta H^0$ of $Ba_7Nb_{3.8}Mo_{1.2}O_{20.1}$ is lower than $\Delta H^0$ values of $Ba_7Nb_4MoO_{20}$ and $Ba_7Nb_{3.9}Mo_{1.1}O_{20.05}$ (Table 3.9). The $\Delta S^0$ of $Ba_7Nb_{3.8}Mo_{1.2}O_{20.1}$ is also lower than $\Delta S^0$ values of $Ba_7Nb_4MoO_{20}$ and $Ba_7Nb_{3.9}Mo_{1.1}O_{20.05}$ (Table 3.9).

**Table 3.9** Hydration enthalpy and entropy of $Ba_7Nb_{4-x}Mo_{1+x}O_{20+x/2} \bullet y$ $H_2O$ ($x = 0, 0.1, 0.2$)

| | $\Delta H^0$/kJ mol$^{-1}$ | $\Delta S^0$/J K$^{-1}$ mol$^{-1}$ | Temperature range/°C |
|---|---|---|---|
| $Ba_7Nb_4MoO_{20}$ | – 22.3 | – 25.7 | 300–400 |
| $Ba_7Nb_{3.9}Mo_{1.1}O_{20.05}$ | – 23.4 | – 24.7 | 300–400 |
| $Ba_7Nb_{3.8}Mo_{1.2}O_{20.1}$ | – 26.7 | – 29.9 | 300–400 |

## 3.4 Conclusion

In this work, we have discovered a new dual-ion (oxide-ion and proton) conducting hexagonal perovskite-related oxide $Ba_7Nb_{3.8}Mo_{1.2}O_{20.1}\cdot y\ H_2O$ by synthesizing the $Ba_7Nb_{4-x}Mo_{1+x}O_{20+x/2}\cdot y\ H_2O$ samples and systematically investigating their electrical and structural properties. $Ba_7Nb_{3.8}Mo_{1.2}O_{20.1}$ exhibits higher bulk conductivity in dry atmospheres compared with known $Ba_7Nb_4MoO_{20}$-based oxides, LSGM [49] below 511 °C, YSZ [47] below 593 °C, and even $(Bi_{0.88}Dy_{0.08}W_{0.04})_2O_{3.12}$ [46] below 357 °C. The origin of the high oxide-ion conductivity at intermediate and low temperatures in $Ba_7Nb_{4-x}Mo_{1+x}O_{20+x/2}$ is the low activation energy for conductivity. The present work has demonstrated that the high conductivity and low activation energy of $Ba_7Nb_{4-x}Mo_{1+x}O_{20+x/2}$ are ascribed to (i) concerted migration via the breaking and reforming of $M_2O_9$ dimers, (ii) high carrier concentration (large amounts of $M_2O_9$ dimers, interstitial oxygen O5 atom and excess oxygen atom $O_{x/2}$ in $Ba_7Nb_{4-x}Mo_{1+x}O_{20+x/2}$) leading to higher minimum NSLD on the O1–O5 path, (iii) short O1–O5 paths across the unit cell, and (iv) the long (Nb/Mo2)–Ba2 distance sandwiching the oxide-ion conducting c′ layer. These factors are facilitated by the unique, intrinsically oxygen-deficient c′ layer in the hexagonal perovskite-related oxide $Ba_7Nb_{3.8}Mo_{1.2}O_{20.1}$. The breaking and reforming of $M_2O_9$ dimers and $MO_5$ monomers are enabled by the flexible coordination of the Nb and Mo cations. Thanks to these factors, the oxide-ion conductivity of $Ba_7Nb_{3.8}Mo_{1.2}O_{20.1}$ is the highest among those of known oxides exhibiting oxide-ion conduction via breaking and reforming of dimers. This work has also demonstrated that protons migrate not only in the oxygen-deficient c′ layer but also in the h layers with fully occupied oxygen sites. It has been suggested that the protons in the c′ layer are mainly located near the corner-sharing oxygen atoms of the $M_2O_9$ dimer, indicating the importance of the $M_2O_9$ dimer for the hydration and migration of protons in the c′ layer. $Ba_7Nb_{3.8}Mo_{1.2}O_{20.1}\cdot y\ H_2O$ exhibits higher total (oxide-ion and proton) and proton conductivities in wet air compared with the mother material $Ba_7Nb_4MoO_{20}\cdot y\ H_2O$. The present findings of high oxide-ion, proton and dual-ion (total oxide ion and proton) conduction and unique migration mechanisms in $Ba_7Nb_{3.8}Mo_{1.2}O_{20.1}\cdot y\ H_2O$ would develop the science and engineering of oxide-ion, proton and dual-ion (total oxide ion and proton) conductors. Future work on $Ba_7Nb_{4-x}Mo_{1+x}O_{20+x/2}\cdot y\ H_2O$ should include the investigation of local structures by total neutron and X-ray scattering and NMR.

## References

1. Staffell I, Scamman D, Abad AV, Balcombe P, Dodds PE, Ekins P, Shah N, Ward KR (2019) The role of hydrogen and fuel cells in the global energy system. Energy Environ Sci 12:463–491
2. Coduri M, Karlsson M, Malavasi L (2022) Structure-property correlation in oxide-ion and proton conductors for clean energy applications: recent experimental and computational advancements. J Mater Chem A 10:5082–5110

3. Zhang W, Yashima M (2022) Recent developments in oxide ion conductors: focusing on Dion-Jacobson phases. Chem Commun 59:134–152
4. Kikuchi Y, Yasui Y, Hester JR, Yashima M (2023) Improved oxide ion conductivity of hexagonal perovskite-related oxides $Ba_3W_{1+x}V_{1-x}O_{8.5+x/2}$. Inorganics 11(6):238
5. Yaguchi H, Morikawa D, Saito T, Tsuda K, Yashima M, Yaguchi H, Yashima M, Morikawa D, Saito T, Tsuda K (2023) High oxide-ion conductivity through the interstitial oxygen site in sillén oxychlorides. Adv Funct Mater:2214082
6. Morikawa R, Murakami T, Fujii K, Avdeev M, Ikeda Y, Nambu Y, Yashima M (2023) High proton conduction in $Ba_2LuAlO_5$ with highly oxygen-deficient layers. Commun Mater 4(1):1–9
7. Shi H, Su C, Ran R, Cao J, Shao Z (2020) Electrolyte materials for intermediate-temperature solid oxide fuel cells. Prog Nat Sci 30(6):764–774
8. Rauf S, Zhu B, Yousaf Shah MAK, Tayyab Z, Attique S, Ali N, Mushtaq N, Wang B, Yang C, Asghar MI, Lund PD (2020) Application of a triple-conducting heterostructure electrolyte of $Ba_{0.5}Sr_{0.5}Co_{0.1}Fe_{0.7}Zr_{0.1}Y_{0.1}O_3$ and $Ca_{0.04}Ce_{0.80}Sm_{0.16}O_2$ in a high-performance low-temperature solid oxide fuel cell. ACS Appl Mater Interfaces 12(31):35071–35080
9. Sažinas R, Bernuy-López C, Einarsrud MA, Grande T (2016) Effect of $CO_2$ exposure on the chemical stability and mechanical properties of $BaZrO_3$-ceramics. J Am Ceram 99(11):3685–3695
10. Murakami T, Hester JR, Yashima M (2020) High proton conductivity in $Ba_5Er_2Al_2ZrO_{13}$, a hexagonal perovskite-related oxide with intrinsically oxygen-deficient layers. J Am Chem Soc 142(27):11653–11657
11. Yang L, Wang S, Blinn K, Liu M, Liu Z, Cheng Z, Liu M (2009) Enhanced sulphur and coking tolerance of a mixed ion conductor for SOFCs: $BaZr_{0.1}Ce_{0.7}Y_{0.2-x}Yb_xO_{3-\delta}$. Science 326(5949):126–129
12. Zhou C, Sunarso J, Song Y, Dai J, Zhang J, Gu B, Zhou W, Shao Z (2019) New reduced-temperature ceramic fuel cells with dual-ion conducting electrolyte and triple-conducting double perovskite cathode. J Mater Chem A 7:13265–13274
13. Fop S, McCombie KS, Wildman EJ, Skakle JMS, Irvine JTS, Connor PA, Savaniu C, Ritter C, Mclaughlin AC (2020) High oxide ion and proton conductivity in a disordered hexagonal perovskite. Nat Mater 19(7):752–757
14. Cao J, Su C, Ji Y, Yang G, Shao Z (2021) Recent advances and perspectives of fluorite and perovskite-based dual-ion conducting solid oxide fuel cells. J Energy Chem 57:406–427
15. Yashima M, Tsujiguchi T, Sakuda Y, Yasui Y, Zhou Y, Fujii K, Torii S, Kamiyama T, Skinner SJ (2021) High oxide-ion conductivity through the interstitial oxygen site in $Ba_7Nb_4MoO_{20}$-based hexagonal perovskite related oxides. Nat Commun 12:556
16. Murakami T, Shibata T, Yasui Y, Fujii K, Hester JR, Yashima M (2021) High oxide-ion conductivity in a hexagonal perovskite-related oxide $Ba_7Ta_{3.7}Mo_{1.3}O_{20.15}$ with cation site preference and interstitial oxide ions. Small 18(10):2106785
17. Sakuda Y, Hester JR, Yashima M (2022) Improved oxide-ion and lower proton conduction of hexagonal perovskite-related oxides based on $Ba_7Nb_4MoO_{20}$ by $Cr^{6+}$ doping. J Ceram Soc Jpn 130(7):442–447
18. Suzuki Y, Murakami T, Fujii K, Hester JR, Yasui Y, Yashima M (2022) Simultaneous reduction of proton conductivity and enhancement of oxide-ion conductivity by aliovalent doping in $Ba_7Nb_4MoO_{20}$. Inorg Chem 61(19):7537–7545
19. Murakami T, Avdeev M, Morikawa R, Hester JR, Yashima M, Murakami T, Avdeev M, Morikawa R, Yashima M, Hester JR (2023) High proton conductivity in β-$Ba_2ScAlO_5$ enabled by octahedral and intrinsically oxygen-deficient layers. Adv Funct Mater 33(7):2206777
20. Nakayama S (2022) Characteristics of electrolyte-supported SOFC single cells using hexagonal perovskite $Ba_3MoNbO_{8.5}$ and $Ba_7Nb_4MoO_{20}$ ceramics. Results Mater 16:100318
21. Bezdorozhev O, Solodkyi I, Ostroverkh A, Morozov I, Ostroverkh Y, Solonin Y (2022) Synthesis and characterization of $(Ba_{1-x}Sr_x)_7Nb_4MoO_{20}$ powders for proton-conducting solid oxide fuel cells. Ukr Chem J 88(4):63–78

22. Yashima M, Yamada H, Nuansaeng S, Ishihara T (2012) Role of $Ga^{3+}$ and $Cu^{2+}$ in the high interstitial oxide-ion diffusivity of $Pr_2NiO_4$-based oxides: design concept of interstitial ion conductors through the higher-valence $d^{10}$ dopant and Jahn-Teller effect. Chem Mater 24(21):4100–4113
23. Kuang X, Green MA, Niu H, Zajdel P, Dickinson C, Claridge JB, Jantsky L, Rosseinsky MJ (2008) Interstitial oxide ion conductivity in the layered tetrahedral network melilite structure. Nat Mater 7(6):498–504
24. Li J, Pan F, Geng S, Lin C, Palatinus L, Allix M, Kuang X, Lin J, Sun J (2020) Modulated structure determination and ion transport mechanism of oxide-ion conductor $CeNbO_{4+\delta}$. Nat Commun 11(1):1–9
25. Fop S, Dawson JA, Fortes AD, Ritter C, McLaughlin AC (2021) Hydration and ionic conduction mechanisms of hexagonal perovskite derivatives. Chem Mater 33(12):4651–4660
26. Kendrick E, Kendrick J, Knight KS, Islam MS, Slater PR (2007) Cooperative mechanisms of fast-ion conduction in gallium-based oxides with tetrahedral moieties. Nat Mater 6(11):871–875
27. Izumi F, Momma K (2007) Three-dimensional visualization in powder diffraction. Solid State Phenom 130:15–20
28. Schönleber M, Klotz D, Ivers-Tiffée E (2014) A method for improving the robustness of linear Kramers-Kronig validity tests. Electrochim Acta 131:20–27
29. Torii S, Yonemura M, Surya Panca Putra TY, Zhang J, Miao P, Muroya T, Tomiyasu R, Morishima T, Sato S, Sagehashi H, Noda Y, Kamiyama T (2011) Super high resolution powder diffractometer at J-PARC. J Phys Soc 80(Suppl B):20
30. Torii S, Yonemura M, Ishikawa Y, Miao P, Tomiyasu R, Satoh S, Noda Y, Kamiyama T (2014) Improvement of instrument devices for super high resolution powder diffractometer at J-PARC. J Phys Conf Ser 502(1):012052
31. Yonemura M, Mori K, Kamiyama T, Fukunaga T, Torii S, Nagao M, Ishikawa Y, Onodera Y, Adipranoto DS, Arai H, Uchimoto Y, Ogumi Z (2014) Development of SPICA, new dedicated neutron powder diffractometer for battery studies. J Phys Conf Ser 502(1):012053
32. Avdeev M, Hester JR (2018) ECHIDNA: a decade of high-resolution neutron powder diffraction at OPAL. J Appl Crystallogr 51(6):1597–1604
33. Ohoyama K, Kanouchi T, Nemoto K, Ohashi M, Kajitani T, Yamaguchi Y (1998) The new neutron powder diffractometer with a multi-detector system for high-efficiency and high-resolution measurements. Jpn J Appl Phys 37:3319–3326
34. Oishi R, Yonemura M, Nishimaki Y, Torii S, Hoshikawa A, Ishigaki T, Morishima T, Mori K, Kamiyama T (2009) Rietveld analysis software for J-PARC. Nucl Instrum Methods Phys Res A 600(1):94–96
35. Oishi-Tomiyasu R, Yonemura M, Morishima T (2012) Application of matrix decomposition algorithms for singular matrices to the Pawley method in Z-Rietveld. J Appl Crystallogr 45(2):299–308
36. Ishikawa Y, Zhang J, Kiyanagi R, Yonemura M, Matsukawa T, Hoshikawa A, Ishigaki T, Torii S, Oishi-Tomiyasu R, Kamiyama T (2018) Z-MEM, maximum entropy method software for electron/nuclear density distribution in Z-code. Phys B: Condens Matter 551:472–475
37. Momma K, Ikeda T, Belik AA, Izumi F (2013) Dysnomia, a computer program for maximum-entropy method (MEM) analysis and its performance in the MEM-based pattern fitting. Powder Diffr 28(3):184–193
38. Adams S (2000) Modelling ion conduction pathways by bond valence pseudopotential maps. Solid State Ion 136–137:1351–1361
39. Chen H, Wong LL, Adams S (2019) SoftBV—a software tool for screening the materials genome of inorganic fast ion conductors. Acta Crystallogr B Struct Sci Cryst Eng Mater 75(1):18–33
40. Momma K, Izumi F (2011) VESTA 3 for three-dimensional visualization of crystal, volumetric and morphology data. J Appl Crystallogr 44(6):1272–1276
41. García-González E, Parras M, González-Calbet JM (1999) Crystal structure of an unusual polytype: 7H-$Ba_7Nb_4MoO_{20}$. Chem Mater 11(2):433–437

42. Shannon RD (1976) Revised effective ionic radii and systematic studies of interatomic distances in halides and chalcogenides. Acta Crystallogr Sect A 32(5):751–767
43. Wegmann M, Watson L, Hendry A (2004) XPS analysis of submicrometer barium titanate powder. J Am Ceram Soc 87(3):371–377
44. Choi JG, Thompson LT (1996) XPS study of As-prepared and reduced molybdenum oxides. Appl Surf Sci 93(2):143–149
45. Yu J, Li Y, Xu X, Duan G, Li Y, Zhou W (2021) Rambutan-like $Nb_2O_5$@SHCs microspheres for improved microwave absorption performance. Compos Commun 24:100643
46. Jung DW, Duncan KL, Wachsman ED (2010) Effect of total dopant concentration and dopant ratio on conductivity of $(DyO_{1.5})_x$-$(WO_3)_y$-$(BiO_{1.5})_{1-x-y}$. Acta Mater 58(2):355–363
47. Kwon OH, Choi GM (2006) Electrical conductivity of thick film YSZ. Solid State Ion 177(35–36):3057–3062
48. Huang K, Tichy RS, Goodenough JB (1998) Superior perovskite oxide-ion conductor; strontium- and magnesium-doped $LaGaO_3$: I, Phase relationships and electrical properties. J Am Ceram Soc 81(10):2565–2575
49. Huang K, Feng M, Goodenough JB (1998) Synthesis and electrical properties of dense $Ce_{0.9}Gd_{0.1}O_{1.95}$ ceramics. J Am Ceram Soc 81(2):357–362
50. Yang X, Zeng X, Ming X, Yang L, Fernández-Carrión AJ, Deng S, He L, Kuang X (2022) Oxide-ion conductivity optimization in $BiVO_4$ Scheelite by an acceptor doping strategy. Inorg Chem Front 9(11):2644–2658
51. Fop S, Dawson JA, Tawse DN, Skellern MG, Skakle JMS, McLaughlin AC (2022) Proton and oxide ion conductivity in Palmierite oxides. Chem Mater 34(18):8190–8197
52. Fop S, Skakle JMS, McLaughlin AC, Connor PA, Irvine JTS, Smith RI, Wildman EJ (2016) Oxide ion conductivity in the hexagonal perovskite derivative $Ba_3MoNbO_{8.5}$. J Am Chem Soc 138(51):16764–16769
53. Lacorre P, Goutenoire F, Bohnke O, Retoux R, Lallgant Y (2000) Designing fast oxide-ion conductors based on $La_2Mo_2O_9$. Nature 404(6780):856–858
54. Zhang W, Fujii K, Niwa E, Hagihala M, Kamiyama T, Yashima M (2020) Oxide-ion conduction in the Dion-Jacobson phase $CsBi_2Ti_2NbO_{10-\delta}$. Nat Commun 11(1):1224
55. Nakayama S, Sakamoto M (1998) Electrical properties of new type high oxide ionic conductor $RE_{10}Si_6O_{27}$ (RE = La, Pr, Nd, Sm, Gd, Dy). J Eur Ceram Soc 18(10):1413–1418
56. Esaka T, Mina-ai T, Iwahara H (1992) Oxide ion conduction in the solid solution based on the Scheelite-type oxide $PbWO_4$. Solid State Ion 57(3–4):319–325
57. Yang X, Fernández-Carrión AJ, Wang J, Porcher F, Fayon F, Allix M, Kuang X (2018) Cooperative mechanisms of oxygen vacancy stabilization and migration in the isolated tetrahedral anion Scheelite structure. Nat Commun 9(1):4484
58. Geng S, Fernández-Carrión AJ, Xu J, Yi H, Lv Y, Kuang X (2020) Electrical properties and oxide ion conducting mechanism in Na-doped $LaPO_4$. Scr Mater 178:527–532
59. Li J, Yang L, Xu J, Fernández-Carrión AJ, Kuang X (2022) Tetrahedral tilting and oxygen vacancy stabilization and migration in $La_{1-x}Sr_{2+x}(GaO_4)O_{1-0.5x}$ mixed electronic/oxide ionic conductors. Inorg Chem 61:5413–5424
60. Packer RJ, Skinner SJ, Yaremchenko AA, Tsipis EV, Kharton VV, Patrakeev MV, Bakhteeva YA (2006) Lanthanum substituted $CeNbO_{4+\delta}$ Scheelites: mixed conductivity and structure at elevated temperatures. J Mater Chem 16(34):3503–3511
61. Li S, Schönberger F, Slater P (2003) $La_{1-x}Ba_{1+x}GaO_{4-x/2}$: a novel high temperature proton conductor. Chem Commun 21:2694–2695
62. Noirault S, Célérier S, Joubert O, Caldes MT, Piffard Y (2007) Incorporation of water and fast proton conduction in the inherently oxygen-deficient compound $La_{26}O_{27}\square(BO_3)_8$. Adv Mater 19(6):867–870
63. López-Vergara A, Porras-Vázquez JM, Infantes-Molina A, Canales-Vázquez J, Cabeza A, Losilla ER, Marrero-López D (2017) Effect of preparation conditions on the polymorphism and transport properties of $La_{6-x}MoO_{12-\delta}$ $(0 \le x \le 0.8)$. Chem Mater 29(16):6966–6975
64. Huse M, Norby T, Haugsrud R (2011) Proton conductivity in acceptor-doped $LaVO_4$. J Electrochem Soc 158(8):B857

65. Fuller CA, Blom DA, Vogt T, Evans IR, Evans JSO (2022) Oxide ion and proton conductivity in a family of highly oxygen-deficient perovskite derivatives. J Am Chem Soc 144(1):615–624
66. Amsif M, Marrero-López D, Magrasó A, Peña-Martínez J, Ruiz-Morales JC, Núñez P (2009) Synthesis and characterisation of $BaCeO_3$-based proton conductors obtained from freeze-dried precursors. J Eur Ceram Soc 29(1):131–138
67. Zhong Z, Li Z, Li J, Guo X, Hu Q, Feng Y, Sun H (2022) A facile method to synthesize $BaZr_{0.1}Ce_{0.7}Y_{0.1}Yb_{0.1}O_{3-\delta}$ (BZCYYb) nanopowders for the application on highly conductive proton-conducting electrolytes. Int J Hydrog Energy 47(94):40054–40066
68. Gagné OC, Hawthorne FC (2015) Comprehensive derivation of bond-valence parameters for ion pairs involving oxygen. Acta Crystallogr B Struct Sci Cryst Eng Mater 71(5):562–578
69. Chambers MS, McCombie KS, Auckett JE, McLaughlin AC, Irvine JTS, Chater PA, Evans JSO, Evans IR (2019) Hexagonal perovskite related oxide ion conductor $Ba_3NbMoO_{8.5}$: phase transition, temperature evolution of the local structure and properties. J Mater Chem A 7(44):25503–25510
70. Yashima M, Tsujiguchi T, Fujii K, Niwa E, Nishioka S, Hester JR, Maeda K (2019) Direct evidence for two-dimensional oxide-ion diffusion in the hexagonal perovskite-related oxide $Ba_3MoNbO_{8.5-\delta}$. J Mater Chem A 7(23):13910–13916
71. Yasui Y, Tsujiguchi T, Sakuda Y, Hester JR, Yashima M (2022) Oxide-ion occupational disorder, diffusion path, and conductivity in hexagonal perovskite derivatives $Ba_3WNbO_{8.5}$ and $Ba_3MoNbO_{8.5}$. J Phys Chem C 126(5):2383–2393
72. Jones A, Slater PR, Islam MS (2008) Local defect structures and ion transport mechanisms in the oxygen-excess apatite $La_{9.67}(SiO_4)_6O_{2.5}$. Chem Mater 20(15):5055–5060
73. Kendrick E, Islam MS, Slater PR (2008) Atomic-scale mechanistic features of oxide ion conduction in apatite-type germanates. Chem Commun 6:715–717
74. Holmes L, Peng L, Heinmaa I, O'Dell LA, Smith ME, Vannier R-N, Grey CP (2008) Variable-temperature $^{17}O$ NMR study of oxygen motion in the anionic conductor $Bi_{26}Mo_{10}O_{69}$. Chem Mater 20(11):3638–3648
75. Gao Y, Nolan AM, Du P, Wu Y, Yang C, Chen Q, Mo Y, Bo S-H (2020) Classical and emerging characterization techniques for investigation of ion transport mechanisms in crystalline fast ionic conductors. Chem Rev 120(13):5954–6008
76. Yashima M, Sirikanda N, Ishihara T (2010) Crystal structure, diffusion path, and oxygen permeability of a $Pr_2NiO_4$-based mixed conductor $(Pr_{0.9}La_{0.1})_2(Ni_{0.74}Cu_{0.21}Ga_{0.05})O_{4+\delta}$. J Am Chem Soc 132(7):2385–2392
77. McCombie KS, Wildman EJ, Fop S, Smith RI, Skakle JMS, McLaughlin AC (2018) The crystal structure and electrical properties of the oxide ion conductor $Ba_3WNbO_{8.5}$. J Mater Chem A 6(13):5290–5295
78. Gilane A, Fop S, Sher F, Smith RI, McLaughlin AC (2020) The relationship between oxide-ion conductivity and cation vacancy order in the hybrid hexagonal perovskite $Ba_3VWO_{8.5}$. J Mater Chem A 8(32):16506–16514
79. Momma K, Izumi F (2019) VESTA: a three-dimensional visualization system for electronic and structural analysis
80. Yasui Y, Tansho M, Fujii K, Sakuda Y, Goto A, Ohki S, Mogami Y, Iijima T, Kobayashi S, Kawaguchi S, Osaka K, Ikeda K, Otomo T, Yashima M (2023) Hidden chemical order in disordered $Ba_7Nb_4MoO_{20}$ revealed by resonant X-ray diffraction and solid-state NMR. Nat Commun 14:2337
81. Novak A (2007) Hydrogen bonding in solids correlation of spectroscopic and crystallographic data. In: Large molecules, pp 177–216
82. Boukamp BA (1995) A linear Kronig-Kramers transform test for immittance data validation. J Electrochem Soc 142(6):1885–1894
83. Boukamp B (2004) Electrochemical impedance spectroscopy in solid state ionics: recent advances. Solid State Ion 169(1–4):65–73

# Chapter 4
# Summary

## 4.1 Overview and Potential of This Thesis

Hexagonal perovskite-related oxide $Ba_7Nb_4MoO_{20}$-based materials have the intrinsically oxygen deficiency layer, and are dual ion conductors that exhibit high oxide ion conductivity and proton conductivity. In this thesis, by substitution for $Ba_7Nb_4MoO_{20}$, I discovered new materials with even higher oxide-ion conductivity and dual-ion conductivity, and clarified the reasons that led to this.

In Chap. 2, I substituted part of the Nb for $Ba_7Nb_4MoO_{20}$ with Cr to synthesize new materials $Ba_7Nb_{4-x}Cr_xMoO_{20+x/2}$ ($x = 0.05, 0.1, 0.15, 0.2, 0.25, 0.3, 0.35, 0.4, 0.45, 0.5$). I measured the conductivity of these materials using a direct current 4-probe method in air and discovered that $Ba_7Nb_{3.8}Cr_{0.2}MoO_{20.1}$ exhibits the highest conductivity among $Ba_7Nb_{4-x}Cr_xMoO_{20+x/2}$. The conductivity of $Ba_7Nb_{3.8}Cr_{0.2}MoO_{20.1}$ was measured in dry gas while changing the oxygen partial pressure by the DC 4-probe method, and it was shown that the conductivity at 603 °C is a highly stable material that is independent of the oxygen partial pressure. Furthermore, the electrical conductivity of $Ba_7Nb_{3.8}Cr_{0.2}MoO_{20.1}$ by DC 4-probe method in wet and dry air revealed that the material exhibits low proton conductivity. By substituting part of the Nb for $Ba_7Nb_4MoO_{20}$ with Cr, I improved the oxide ion conductivity and suppressed the proton conduction simultaneously. Rietveld and MEM analyses were performed by neutron diffraction data of $Ba_7Nb_{3.8}Cr_{0.2}MoO_{20.1}$ at high temperatures. Oxide ions two-dimensionally migrate through the lattice O1 and interstitial O5 sites in the c' layer via the interstitialcy diffusion mechanism. To determine the site preference of Cr, preliminary analyses were performed to compare the reliability factors for the structural model where Cr was placed in $M1$, $M2$ and $M3$, and it was shown that Cr was located near the c′ layer. In future, $Ba_7Nb_{3.8}Cr_{0.2}MoO_{20.1}$ is expected to be developed as an electrolyte material for SOFCs because it shows high oxide-ion conductivity, low proton conductivity, and wide electrolyte range. Furthermore, since AIMD simulations of $Ba_7Nb_{3.8}Cr_{0.2}MoO_{20.1}$ have not yet been performed, we will



Y. Sakuda, *Discovery of $Ba_7Nb_4MoO_{20}$-Based Materials and the Mechanism of Ultrafast Ion Conduction via Dimers*, Springer Theses,
https://doi.org/10.1007/978-981-96-2552-9_4

investigate the migration mechanism of oxide ion in detail focusing on the coordination number of Cr, and we will be able to understand the migration mechanism further.

In Chap. 3, I substituted part of the Nb for $Ba_7Nb_4MoO_{20}$ with Mo to synthesize the new materials $Ba_7Nb_{4-x}Mo_{1+x}O_{20+x/2}$ ($x = 0.02, 0.04, 0.06, 0.08, 0.12, 0.14, 0.16,$ $0.18, 0.2, 0.22, 0.25, 0.3$) to measure their conductivity by DC 4-probe method in static air, and discovered the $Ba_7Nb_{3.8}Mo_{1.2}O_{20.1}$ which shows the highest conductivity highest conductivity among $Ba_7Nb_{4-x}Mo_{1+x}O_{20+x/2}$. By AC measurement, the bulk conductivity of $Ba_7Nb_{3.8}Mo_{1.2}O_{20.1}$ is 1.1 mS $cm^{-1}$ at 306 °C, which is 175 times higher than that of the practical material 8YSZ. Next, for $Ba_7Nb_{3.8}Mo_{1.2}O_{20.1}$, the electrical conductivity of was measured in dry gas while changing the oxygen partial pressure by DC 4-probe method, and the electromotive force of the oxygen concentration cell was measured. The electrical conductivity of $Ba_7Nb_{3.8}Mo_{1.2}O_{20.1}$ was independent of oxygen partial pressure at 302 and 603 °C, and the transport number of the oxide ions was close to 1 at 596–903 °C. $Ba_7Nb_{3.8}Mo_{1.2}O_{20.1}$ shows high oxide-ion conductivity at low temperatures and high stability at low oxygen partial pressure, which is expected to be a new material for SOFCs. The results of structure analyses of $Ba_7Nb_4MoO_{20}$ and $Ba_7Nb_{3.8}Mo_{1.2}O_{20.1}$ by neutron diffraction data at high temperatures showed that $Ba_7Nb_{3.8}Mo_{1.2}O_{20.1}$ has intrinsically oxygen deficiency layer in which oxide ions diffuse 2-dimensionally. The amount of interstitial oxygen in $Ba_7Nb_{3.8}Mo_{1.2}O_{20.1}$ is larger than that in the parent material $Ba_7Nb_4MoO_{20}$, and the minimum NSLD of $Ba_7Nb_{3.8}Mo_{1.2}O_{20.1}$ is higher than that in the parent material $Ba_7Nb_4MoO_{20}$. $Ba_7Nb_4MoO_{20}$-based materials show high oxide-ion conductivity at low temperatures, and this reason is that the distance between the cations sandwiching the c′ layer is long. I believe this is because the charge in the c ’ layer is difficult to be negatively charged, and the c’ layer difficult to attract cations. The dynamic and local structure of the $(Ba_7Nb_4MoO_{20})_9$ and $(Ba_7Nb_{3.8}Mo_{1.2}O_{20.1})_9$ were investigated by AIMD simulations. Oxide ions in $(Ba_7Nb_{3.8}Mo_{1.2}O_{20.1})_9$ diffuse involving the breaking and reforming of $M_2O_9$ dimers, $MO_5$ monomers, and $MO_4$ tetrahedra. These results are believed to be the reason why $Ba_7Nb_{3.8}Mo_{1.2}O_{20.1}$ exhibits high oxide-ion conductivity. Since $Ba_7Nb_{3.8}Mo_{1.2}O_{20.1}$ also exhibits proton conductivity, AIMD simulation of $(Ba_7Nb_{3.8}Mo_{1.2}O_{20.1}{\cdot}0.1\ H_2O)_9$ was performed. Protons migrate from the intrinsically oxygen deficient layer to the hexagonal close-packed layer, and protons rapid diffuse in the hexagonal close-packed layer.

The research in Chap. 3 , I discovered $Ba_7Nb_{3.8}Mo_{1.2}O_{20.1}$ which is a promising material for SOFCs, and clarified the detailed mechanisms for oxide-ion and proton migration in hexagonal perovskite-related oxides. Furthermore, I also discovered a new strategy for improving dual-ion conductivity by simultaneously improving both oxide-ion conductivity and proton conductivity. In future, $Ba_7Nb_{3.8}Mo_{1.2}O_{20.1}$ is expected to be developed as an electrolyte material for SOFCs because it exhibits not only high oxide ion conductivity but also high chemical and electrical stability. $Ba_7Nb_{3.8}Mo_{1.2}O_{20.1}$ exhibits high oxide ion conductivity even at low temperatures, so as thin-film technology advances, it is expected to be able to be used at even lower temperatures than conventional SOFCs. AIMD simulations revealed that oxide-ion migration in $(Ba_7Nb_{3.8}Mo_{1.2}O_{20.1})_9$ is reminiscent of concerted push–pull interstitialcy "bucket-relay" type motion involving the breaking and reforming of $M_2O_9$ dimers, $MO_5$ monomers, and $MO_4$ tetrahedra. The discovery of this motion is expected to lead to similar considerations for other oxides with intrinsically oxygen vacancy layers. I showed that the longer the distance between cations sandwiching the c′ layer, the lower activation energy for oxide ion conductivity. Therefore, I proposed that searching for materials with long distance between cations sandwiching the c′ layer will lead to the discovery of materials that exhibit higher oxide ion conductivity. In addition, $Ba_7Nb_{3.8}Mo_{1.2}O_{20.1}$ is a material that exhibits high dual-ion conductivity, it is expected to be developed as an electrolyte material for fuel cells which both oxide-ion and proton diffuse.

In this work, since a new design guideline has been added for hexagonal perovskite-related oxides with intrinsically oxygen deficiency layer, it is expected to advance the study for electrolyte materials for fuel cells.

The manufacturer's authorised representative in the EU is Springer Nature Customer Service Centre GmbH, Europaplatz 3, 69115 Heidelberg, Germany. If you have any concerns regarding our products, please contact ProductSafety@springernature.com

Printed and bound by CPI Group (UK) Ltd, Croydon, CR0 4YY
01/10/2026
02225945-0005